Samuel O. Idowu · Henk J. de Vries ·
Ivana Mijatovic · Donggeun Choi
Editors

Sustainable Development

Knowledge and Education About Standardisation

Editors
Samuel O. Idowu
Guildhall Faculty of Business and Law
London Metropolitan University
London, UK

Henk J. de Vries
Rotterdam School of Management
Erasmus University
Rotterdam, The Netherlands

Ivana Mijatovic
Faculty of Organizational Science
University of Belgrade
Belgrade, Serbia

Faculty of Technology, Policy
and Management
Delft University of Technology
Delft, The Netherlands

Donggeun Choi
Korean Standards Association
Seoul, Korea (Republic of)

ISSN 2196-7075   ISSN 2196-7083   (electronic)
CSR, Sustainability, Ethics & Governance
ISBN 978-3-030-28714-6   ISBN 978-3-030-28715-3   (eBook)
https://doi.org/10.1007/978-3-030-28715-3

© Springer Nature Switzerland AG 2020
This work is subject to copyright. All rights are reserved by the Publisher, whether the whole or part of the material is concerned, specifically the rights of translation, reprinting, reuse of illustrations, recitation, broadcasting, reproduction on microfilms or in any other physical way, and transmission or information storage and retrieval, electronic adaptation, computer software, or by similar or dissimilar methodology now known or hereafter developed.
The use of general descriptive names, registered names, trademarks, service marks, etc. in this publication does not imply, even in the absence of a specific statement, that such names are exempt from the relevant protective laws and regulations and therefore free for general use.
The publisher, the authors and the editors are safe to assume that the advice and information in this book are believed to be true and accurate at the date of publication. Neither the publisher nor the authors or the editors give a warranty, expressed or implied, with respect to the material contained herein or for any errors or omissions that may have been made. The publisher remains neutral with regard to jurisdictional claims in published maps and institutional affiliations.

This Springer imprint is published by the registered company Springer Nature Switzerland AG
The registered company address is: Gewerbestrasse 11, 6330 Cham, Switzerland

Foreword

Education is a socially constructed phenomenon. And as a process core to what happens in societies, it plays a crucial role in socializing human beings into the norms and values surrounding behavior, individually and as individuals interact in groups. Notably in the UK, policy makers at the time of writing have drawn attention to an over-emphasis on a technocratic view of education and measures of success, encouraging educators to pay attention to a more comprehensive frame of reference for evaluating this important activity.

Norms and values are at the heart of what standardization may involve in education. It is important, by the way, to disentangle education from training, although the two may reasonably be viewed as complementary activities. Training enables skills development; education potentially transforms people's outlooks and modes of behaving—as it enables them to understand what their induction into the human race has involved to the point of conscious reflection on these experiences. And an aspiration to move them forward in particular ways—not only in terms of general education but building on that in following specialist pathways.

Achieving and maintaining practices across educational institutions leading to sustainability in the world they serve depend on public warrant. Being a legitimate process based on certain commonly accepted components. Monitoring and evaluating such components' evolution in practice can benefit from standardization to make common norms available to facilitate these processes, albeit with the caveat to avoid the myopic thinking referred in the opening paragraph. Therefore, mitigating risks that standardization may be harmful; in particular, if it lacks transparency in surfacing the norms and values derived from the interest groups (including education "professionals") involved in determining it.

This recognizes that standardization in education (as in other aspects of social life), and its association with arguments about sustainability in terms of maintaining consent and resources to continue educational processes, is contested terrain. Which is why it is valuable to have a volume such as the present one to bring together systematically for critical imbibing and appraisal the debates contemporarily taking place in the field.

In their introduction, the editors set the scene by reference to "the art of teaching" and, consistent with this enshrining feature, indicate the strategic choice that has been made for contributions to the volume largely to focus bottom-up on ideas and practice. While socialization can be a top-down process, if one considers the institutional processes involved as an art form, then it logically follows to place weight on what emerges from agency and interactions between participants. From which emergent standards may be identified to complement any values-based standards identified in a more deontological manner. As contributors point out, education as *communication* between social actors to enable shared understanding around issues in, e.g., the legal, political, and technological spheres, across the sub-strata of socioeconomic life is needed. In turn, arriving at a standardized basis for agency. Underpinned by attention to what are perceived to be the needs of industry and society, themselves socially constructed phenomena and thus open to contestation.

The contributions to this volume address these challenges and offer readers a wide range of perspectives and focal points from which to engage with the field—elevating attention to its importance in education specifically and for society and industry generally. Not only are the factors delineated and critically appraised, the strategic significance of the terrain is suitably emphasized, along with avenues to understanding and steps to act on such understanding.

January 2019

Stephen J. Perkins, D.Phil. (Oxon)
Emeritus Professor
London Metropolitan University

Senior Research Fellow
Global Policy Institute London

Preface

There is no easy way to address different and specific roles of education and research for sustainable development. Having said this, there is a general consensus that education and research are highly important. Standards and the processes of their development and standardization continue to shape all businesses operating in a globalized world. The impacts of standards development and standards' implementation on economies, societies, and the environment are evident and numerous, albeit, still not fully understood. Many actors, with different intentions and interest, develop standards which can be more or less successful in the market for instance; they might compete in the market for industry acceptance or being referenced in regulations. Engaging in sustainable development issues is not possible without having adequate standards in place. Knowledge and education about standardization can be understood as one of many prerequisites for sustainable development.

Many global actors have called for more education about standardization. In 1970, the UNECE Government Officials Responsible for Standardization Policies (the predecessor of WP. 6) developed recommendations that urged governments to include standardization in the curricular of educational institutions. Almost all standards development organizations (SDOs) have called for education about standardization and activities to support education on standardization. The need for cooperation among industry, academia, and standard organizations is essential in order to promote, initiate, and foster education about standardization in the globalized world.

Despite such initiatives, little attention is paid to the formal education about standardization. The 13 chapters of this book are focused on knowledge and education about standardization in higher education, and this we believe is the first attempt to address these issues in one book. We hope that this book will be valuable to those who are teaching or would like to teach modules on standards and standardization or for those who would like to understand more clearly specific aspects of education on standardization.

London, UK — Samuel O. Idowu
Rotterdam, The Netherlands — Henk J. de Vries
Belgrade, Serbia — Ivana Mijatovic
Seoul, Korea (Republic of) — Donggeun Choi

Acknowledgements

Editing a book is not an easy task but with support from a number of willing scholars, many of the tasks involved are made a lot easier. This was the case with this book. For this reason, these four editors owe a load of thank you to all our contributors who have supported us with their innovative chapters in this book. Despite their busy schedules, they felt obliged to help in putting together this very fine informative addition to the literature on standardization. These four editors are grateful to you all.

We would also like to express our gratitude to a few people starting with Professor Stephen J Perkins, Emeritus Professor, and Senior Research Fellow, London Metropolitan University, who wrote a very fine foreword to the book. We are also grateful to our families and colleagues for their support during the process of editing the book.

The lead editor and the editorial team of the book would like to thank the publishing team at Springer headed by the Executive Editor, Christian Rauscher, Barbara Bethke, and other members of the publishing team who have supported this project and the lead editor's other projects.

Finally, we apologize for any errors or omissions that may appear anywhere in this book, and please be assured that no harm was intended to anybody.

Introduction

Regular readers of this book series may wonder: Why a book on knowledge and education about standardization? Isn't standardization a phenomenon for engineers? Indeed, the vast majority of standards still address products or technology-related processes such as testing. However, the scope of standardization is broadening. The number of standards for services and management systems has been increasing for three decades already, and this increase continues. More recently, standards for complex technical and societal systems are being added. And last but not least, the increased attention for sustainability as reflected in the United Nations Sustainable Development Goals brings sustainability to the company management agenda (Van Tulder, 2018) and brings standards to the societal agenda (De Vries et al., 2018). Standards used to be applied to facilitate for-profit activities and to mitigate negative externalities for, in particular, workers, consumers, and the environment, but the increased attention for societal issues makes that new standards need to be developed taking the starting point at the side of the societal issues. Some standards are available already, and the United Nations Sustainable Development Goals are expected to trigger additional standardization activities (ISO, 2018; Jachia, 2018).

Standardization is and continues to be a major instrument for private goods and services in for-profit market environments. These markets get increasingly intertwined with private non-profit initiatives (e.g., fair trade) and public non-profit initiatives (governmental policies). Common goods are shared between these three, but these suffer from the "tragedy of the commons." Dealing with common pool problems requires the involvement and positive actions of all societal actors: companies, governments, and NGOs. Standards and subsequent conformity assessment are important instruments to avoid a tragedy of the commons, and the involvement of all stakeholders is essential for the development and market acceptance of these standards. These stakeholders need to be well prepared, and this requires education and training. And therefore, this book is needed, in this book series.

Frankly, this book project was a challenge because it aims to bridge three worlds: sustainability, standardization, and education. Most of the papers we received combine standardization and education and are written by academic experts in standardization, experienced in teaching in this field and/or in stimulating education about standardization. Just one paper, written by Wright et al., starts at the sustainability side, and therefore, we decided to start with that one because it moves from sustainability to standards and then to teaching about sustainability standards, using three case studies, on a standard to solve a sustainability issue, carbon footprints, and lifecycle assessment. Also, the next chapters present cases of teaching about standards and standardization: Van den Bosche on standards for electric vehicles, Fomin on standardization for informatics and information systems, and Van de Kaa on battles between competing standards. The last case on teaching, by de Vries, describes a case in which standardization is related to innovation management. Interestingly, he shows that systematic course design makes that sustainability issues are being addressed automatically.

The second part of the book takes a country perspective: Bulgaria (Vasileva) and Romania (Puiu). We intended to have papers about other countries as well, but unfortunately we had to reject several of such papers during the review process.

The third part of the book is about stimulating education about standardization. This starts with the need for it in the market. This is the topic of the first paper, written by Blind and Drechsler. Also, the next paper, by Jachia et al., addresses market needs, by referring to a guide developed by the International Federation of Standards Users IFAN. This guide shows who in a company should know what about standardization. But their paper offers more, and it also provides an overview of initiatives to stimulate education about standardization in different parts of the world and then focuses on the role the United Nations Economic Commission for Europe plays in doing this: by emphasizing the need in the form of a Recommendation to the member states and by means of a model program for education about standardization. Next, De Vries et al. show the role communities of practitioners can play in developing and exchanging standardization knowledge. The author of the next paper, Kanevskaio, discovered that practitioners in standardization may be limited in their view on the topic and then cannot put their standardization activities into perspective. She argues that a multidisciplinary education approach is needed. Then, the challenge is to bring this message also to new generations—Mijatovic shows how "generation Z" can be addressed in education about standardization.

The last chapter discusses the foregoing chapters, draws conclusions, and provides recommendations.

Henk J. de Vries
hvries@rsm.nl

References

De Vries, H. J., Jakobs, K., Egyedi, T., Eto, M., Fertig, S., Klintner, L., … Scaramuzzino, G. (2018). Standardization—Towards an agenda for research. *International Journal of Standardization Research, 16*(1), 52–59.

ISO (2018) *Contributing to the UN Sustainable Development Goals with ISO standards*. Geneva: International Organization for Standardization. https://www.iso.org/files/live/sites/isoorg/files/store/en/PUB100429.pdf.

Jachia, L. (2018). *Standards for the sustainable development goals*. Geneva: United Nations Economic Commission for Europe.

Van Tulder, R. (2018). *Business & the sustainable development goals—A framework for effective corporate involvement*. RSM Series on Positive Change Vol. 0. Rotterdam: Rotterdam School of Management, Erasmus University.

Contents

Cases of Education About Standardization

Standards in the Classroom: A Vehicle for Sustainability Education 3
Laurence A. Wright, Julie Sinistore, Mike Levy and Bill Flanagan

A Look Behind the Curtain: Standardization Education for Engineers from the Electric Vehicle Standardization Shopfloor 17
Peter Van den Bossche

From Boring to Intriguing: Personal Perspective on Making Education About Standardization Appealing to Students 31
Vladislav V. Fomin

Mutual Enforcement of Research and Education—The Case of Structured Inquiry-Based Teaching of Standardization 45
Geerten van de Kaa

Addressing Sustainability in Education About Standardisation—Lessons from the Rotterdam School of Management, Erasmus University 57
Henk J. de Vries

Cases of Countries

Education About Standardization in the Context of Sustainable Development 79
Elka Vasileva

Strategies on Education About Standardization in Romania 95
Silvia Puiu

Stimulating Education About Standardisation

Necessary Competences of Employees in the Field of Standardization 113
Knut Blind and Sandra Drechsler

UNECE Initiatives on Education on Standardization 139
Lorenza Jachia, Serguei Kouzmine and Haiying Xu

The Need for Multi-disciplinary Education About Standardization 161
Olia Kanevskaia

Learning in Communities of Standardisation Professionals 179
Henk J. de Vries, Jeroen Trietsch and Paul M. Wiegmann

Teaching Standardization to Generation Z-Learning Outcomes Define Teaching Methods 191
Ivana Mijatovic

Summary

Sustainable Development: Knowledge and Education About Standardisation—Discussion 211
Henk J. de Vries

Index 215

Editors and Contributors

About the Editors

Samuel O. Idowu, Ph.D. is Senior Lecturer in Accounting and Corporate Social Responsibility at London Guildhall School of Business and Law, London Metropolitan University, UK. He researches in the fields of corporate social responsibility (CSR), corporate governance, business ethics, and accounting and has published in both professional and academic journals since 1989. He is a freeman of the City of London and a Liveryman of the Worshipful Company of Chartered Secretaries and Administrators. He is Deputy CEO and First Vice President of the Global Corporate Governance Institute. He is Editor in Chief of three Springer's reference books—the Encyclopedia of Corporate Social Responsibility, the Dictionary of Corporate Social Responsibility, and the Encyclopedia of Sustainable Management (forthcoming)—he is Editor in Chief of the International Journal of Corporate Social Responsibility (IJCSR), Editor in Chief of the American Journal of Economics and Business Administration (AJEBA), and Associate Editor of the International Journal of Responsible Management in Emerging Economies (IJRMEE). He is also Series Editor for Springer's books on CSR, sustainability, ethics, and governance. One of his edited books won the most Outstanding Business Reference Book Award of the American Library Association (ALA) in 2016 and another was ranked 18th in the 2010 Top 40 Sustainability Books by, *Cambridge University, Sustainability Leadership Programme.* He is Member of the Committee of the Corporate Governance Special Interest Group of the British Academy of Management (BAM). He is on the Editorial Boards of the International Journal of Business Administration, Canada, and Amfiteatru Economic Journal, Romania, and a few more. He has delivered a number of keynote speeches at national and international conferences and workshops on CSR and has on two occasions 2008 and 2014 won Emerald's Highly Commended Literati Network Awards for Excellence. To date, he has edited several books in the field of CSR, sustainability and governance and has written ten forewords to CSR books. He has served as External Examiner to the following UK Universities—Sunderland, Ulster, Anglia

Ruskin, Plymouth, Robert Gordon University, Aberdeen, Teesside University, Middlesbrough, Sheffield Hallam University, and Leicester De Montfort University. He has also examined Ph.D. theses for a few non-UK Universities from across the globe.

Henk J. de Vries (1957) is Professor of Standardisation Management at the Rotterdam School of Management, Erasmus University in Rotterdam, The Netherlands, Department of Technology and Operations Management, Section Innovation Management, and Visiting Professor at Delft University of Technology, Faculty of Technology, Policy and Management, Department of Values, Technology and Innovation, Section Economics of Technology and Innovation. His education and research focus on standardisation from a business point of view. From 1994 until 2003, he worked with NEN, Netherlands Standardization Institute, in several jobs, being responsible for R&D during the last period. Since 1994, he has an appointment at the Erasmus University's School of Management, and since 2004, he has been working full time at this university. He is (co-)author of more than 380 publications on standardization, including several books. See www.rsm.nl/people/henk-de-vries. In 2009, the International Organization for Standardization ISO awarded his education about standardization as best in the world. He is President of the European Academy for Standardization EURAS.

Ivana Mijatovic is Associate Professor at Faculty of Organizational Sciences, University of Belgrade. For much of her academic career, she has focused on standardization and quality management. She is Passionate Teacher; on bachelor studies, she teaches standardization 1, quality engineering and quality planning; on master and Ph.D. studies, she teaches standardization 2 and ICT standardization. In 2018/2019, she is Chair of the Board of the International Cooperation for Education about Standardization ICES (http://www.standards-education.org/). Since 2017, she has been serving as Member of working group related to EU Joint Initiative on Standardization (JIS Action 3). Since 2015, she is Member of the STARTed team (Team of Specialists on Standardization and Regulatory Techniques—education on standardization) of The United Nations Economic Commission for Europe (UNECE). She serves as Vice President on the board of the European Academy for Standardization (EURAS, www.euras.org), and she was Member of the Balkan Coordination Committee for Standardization, Prototypes, and Quality (BCC). She is Member of national technical committee KS I1/07—Software engineering, IT for Education and Internet at Institute for Standardization of Serbia (national mirror committee in relation with ISO/IEC JTC 1/SC 7 Software and systems engineering; CEN/TC 353 Information and Communication Technologies for Learning, Education and Training; CEN/TC 365 Project Committee—Internet Filtering; ISO/IEC JTC 1/SC 36 Information technology for learning, education and training and ISO/IEC JTC 1/SC 40 IT Service Management and IT Governance). She developed the course standardization 1 and wrote a text book with ten case studies standardization 1 (2015). Her current academic work addresses questions of standardization, education about standardization, teaching quality, and quality aspects of technology-enhanced learning.

Donggeun Choi is Chief Researcher in Korean Standards Association (KSA) where he joined in 2000. He holds a Ph.D. in Technology and Innovation Management from Sungkyunkwan University of Korea. He also serves as Secretary of International Cooperation of Education about Standardization (ICES). He is Proponent and Editor of APEC standards education initiative which completed in 2007–2011, and he recently proposed a new APEC project titled "Inspiring Next Generation of Standards Professional Development: Phase I. Identifying Stakeholder Requirements." He has published research articles and policy reports in the area of standardization (governance, patents, and education) and technology innovation.

Contributors

Knut Blind Faculty of Economics and Management, Fraunhofer Institute for Open Communication Systems FOKUS, Technische Universität Berlin, Berlin, Germany

Henk J. de Vries Rotterdam School of Management, Erasmus University, Rotterdam, The Netherlands;
Faculty of Technology, Policy and Management, Delft University of Technology, Delft, The Netherlands

Sandra Drechsler Institut Für Produktentwicklung, Karlsruher Institut Für Technologie (KIT), IPEK, Karlsruhe, Germany

Bill Flanagan Aspire Sustainability, Albany, NY, USA

Vladislav V. Fomin Vilnius University, Kaunas, Lithuania

Lorenza Jachia UNECE Working Party on Trade and Standardization Policies, Economic Cooperation and Trade Division, UNECE, Geneva, Switzerland

Olia Kanevskaia Tilburg Law and Economics Center (TILEC) and Tilburg Law School (TLS), Tilburg, The Netherlands

Serguei Kouzmine UNECE, Geneva, Switzerland

Mike Levy First Environment, Sacramento, CA, USA

Ivana Mijatovic Faculty of Organizational Sciences, University of Belgrade, Belgrade, Serbia

Silvia Puiu Faculty of Economics and Business Administration, University of Craiova, Craiova, Romania

Julie Sinistore WSP, Portland, OR, USA

Jeroen Trietsch Knowledge Network for Continuous Improvement, 's-Hertogenbosch, The Netherlands

Geerten van de Kaa Delft University of Technology, Delft, The Netherlands

Peter Van den Bossche Vrije Universiteit Brussel, Brussel, Belgium

Elka Vasileva University of National and World Economy, Sofia, Bulgaria

Paul M. Wiegmann Technical University of Eindhoven, Eindhoven, The Netherlands

Laurence A. Wright Warsash School of Maritime Science and Engineering, Solent University, Southampton, UK

Haiying Xu Intern, Economic Cooperation and Trade Division, UNECE, Geneva, Switzerland

Cases of Education About Standardization

Standards in the Classroom: A Vehicle for Sustainability Education

Laurence A. Wright, Julie Sinistore, Mike Levy and Bill Flanagan

1 Introduction

The concept of 'Sustainable Development' has been adopted on the agendas of governments, corporations, education, and in research. The United Nations 2005 World Summit described sustainability in terms of the triple bottom line approach—social equity, economic prosperity, and environmental quality. At a global level the UN recently adopted 17 Sustainable Development Goals (SDGs). Signatory nations have committed to tackle deep-rooted issues such as gender inequality climate change, and access to quality education. The goals represent many of the 'wicked' challenges characteristic of sustainable development and require coordinated responses (Rittel & Webber, 1973; United Nations General Assembly, 2015). Government and private sector investment is essential to making progress towards the sustainability outcomes, contributing to job creation, increasing income, providing services, and increased resilience (United Nations General Assembly, 2015). The challenges identified of sustainability are global and we cannot find solutions without suitable and agreed mechanisms. However, the global, long-term, complex nature of has largely meant an absence of agreed inter-governmental regulation. In the absence of inter-

---

L. A. Wright (✉)
Warsash School of Maritime Science and Engineering, Solent University, Southampton, UK
e-mail: laurie.wright@solent.ac.uk

J. Sinistore
WSP, Portland, OR, USA
e-mail: julie.sinistore@wsp.com

M. Levy
First Environment, Sacramento, CA, USA
e-mail: MLevy@FirstEnvironment.com

B. Flanagan
Aspire Sustainability, Albany, NY, USA
e-mail: bill@aspiresustainability.com

© Springer Nature Switzerland AG 2020
S. O. Idowu et al. (eds.), *Sustainable Development*, CSR, Sustainability,
Ethics & Governance, https://doi.org/10.1007/978-3-030-28715-3_1

governmental leadership, private actors, including corporations and NGOs, have become increasing active through standardisation of transnational affairs in pursuit of the promotion of social equity and ecological sustainability objectives.

Sustainability standards are becoming an increasingly important mechanism to drive sustainability objectives. Standards are ubiquitous in modern life; from standards to ensure technological interoperability to the quality of products and the environmental performance of organisations (Brunsson et al., 2012). The most effective standards, of course, go completely unnoticed in our daily lives (e.g. food company recipes, music and video compatibility). Some of the most widely known international sustainability standards systems are well known brands [e.g. the Marine Stewardship Council (MSC), Forestry Stewardship Council (FSC)], with consumers relying on recognition of these brands to inform purchasing decisions (Komives & Jackson, 2014). The International Organization for Standardization (ISO) 14000 series of standards receive significant attention as a market-driven approach to environmental protection, providing an alternative to top-down regulation (Stenzel, 2000). Interestingly although the ISO standards are voluntary one of the key requirements of ISO14001 Environmental Management Systems, is to ensure that businesses comply with all relevant and applicable environmental legislation. Effectively promoting environmental protection measures, whilst providing a means to self-regulate, minimising exposure to surveillance and sanction.

Within this paper we define sustainability standards as simply an agreed way of doing something, such as producing a product or running a process in a manner which promotes good social and/or environmental practices. We acknowledge the limitations of this approach and recognise that there are numerous methods for the classification of standards (e.g. see: Egyedi & Ortt, 2017). Sustainability standards help consumers identify sustainable solutions and products, guide purchasing decisions, and establish agreed sustainable practices in a range of industries (Manning, Boons, Von Hagen, & Reinecke, 2012). These standards can be broadly subdivided into three categories: product (e.g. the quality and safety of goods), process (e.g. processes of manufacture and packaging), and management systems (i.e. organisational management).

While there is general consensus about the generic propositions of 'sustainability', it is contested both as a concept and objective by the various actors involved. Over time this has the effect of promulgation and co-evolution of numerous sustainability standards (Manning et al., 2012). Examples can be found in many industries, for example the coffee industry has in excess of thirty corporate, multi-stakeholder standards developed over the past two decades—e.g. Rainforest Alliance; Fairtrade (Reinecke, Manning, & Von Hagen, 2012). The phenomenon of multiple standards defies established convention that overtime a de facto accepted standard arises where there is enough critical mass among customers and industry to effectively use the standard. Thereby numerous actors or groups thereof co-define the norms and objectives of sustainable development to which the value chain must adhere.

Given the ubiquitous nature of standards and the importance as a mechanism in the pursuit of sustainability objectives, it is vital that graduates entering industry are equipped with the knowledge and skills to interpret and apply these standards.

Students and learners at all levels are the future practitioners, managers, and industry representatives. To assist these students in finding employment in technical specialist roles, the development of a well-defined employability profile with both specialist and generic skills, is of benefit to both the student and the employer (Kemp, 2009). Higher education plays a central role in developing the skills and attributes desired by employers, society, and the economy. This is reflected in the attitudes of students with large numbers reporting enhancements to employability profiles to be one of the primary motivators for attending university (Glover et al., 2002; Kemp et al., 2008). The practice of embedding these employability skills in the curriculum is a subject of debate within the sector (Kemp, 2008). This debate is likely to continue with increasing numbers of student placing greater demand on employability outcomes. The expectation of employers encompasses not only generic employability skills and attributes, but for technical specialist roles, an advanced level of expertise and competence in key subject areas, such as the knowledge and ability to apply and interpret standards (Kemp, 2008).

Developing an understanding of the role and operation of standards is therefore vital for graduates of environmental and sustainability programmes, and more widely throughout education. Whilst students may be formally introduced to standards during their university education, they can develop a fundamental understanding of basic principles of sustainability (and informal consensus standards) throughout their education. This chapter explores the use of simulation pedagogies, with three case studies, to bring both the underlying concepts of sustainability and the knowledge and skills necessary to interpret and apply sustainability standards to the classroom. We argue that the common underlying simulation pedagogy offers a route to bring deep conceptual and theoretical understanding to students.

2 Simulation Pedagogies

Developing knowledge and understanding of standards requires the concurrent development of professional skills, understanding of business context, and subject knowledge. A substantial body of literature suggests that 'non-traditional' techniques and approaches, such as games or simulation, and other interactive approaches may help to involve and engage students with their learning and develop these attributes (Kumar & Lightner, 2007).

The argument for active learning methods are well established. In a traditional lecture environment, the student passively receives information from the instructor and may not find themselves invested in the content. This is compounded where higher education curriculum and classroom environments tend to be more flexibly structured than at other levels of education (e.g. high school); subsequently students often experience difficulties in motivating themselves to pursue learning goals. Making student motivation and engagement are an ongoing challenge for instructors and educators (Glynn, Aultman, & Owens, 2005). Conversely learners required to take an active role learning will find themselves more engaged and benefit from a deeper learning process (Prince, 2004). Techniques to encourage moving around a room, participation in a competition, or discussion with fellow learners can raise student awareness, engagement, and help mitigate learner fatigue (Kumar & Lightner, 2007).

Simulations and games can bridge the gap between the classroom and real life. Encouraging active student participation and developing a sense of understanding of real-world application. Participation in these activities requires the use and application of taught content by the learner; therefore, ensuring students are achieving a deeper level of engagement with the content provided to them. A well-executed game views the learner as the most important actor in the learning process, the learner has to choose the route(s) in the creation and application of knowledge (Bíró, 2014). Learners may compete against each other to reach objectives or 'higher levels' of the game, perhaps earning rewards (e.g. badges, points). Alternatively the simulation may simply present a safe space for experimentation and exploration without the pressures of assessment. Simulations serve to 'bring material to life' without sacrificing the content or context of the learning. Much like card games or sports, standards make sense when they are applied, or in this context 'played' (Asal & Blake, 2006). Simulation exercises enable students to experience and experiment with standards much as the same way that a laboratory experiment allows students to observe and apply physical processes. Dieleman and Huisingh (2006) advocate experiential learning in sustainable development education to address the challenges of systems thinking, collaboration, and active learning.

Simulation pedagogies may be comparatively simple—for example the classical prisoner's dilemma can be represented very simply. Or further depth and complexity may be added for example to create a global political negotiation, which requires a more in-depth and complex structure (Asal & Blake, 2006; Kemp et al., 2015). Role-playing exercises [e.g. the UN Model Nations (UNA-UK, 2016)] provide established, often extra-curricular activities, for students to enhance their understanding of global affairs, and develop their skills (Horn, Rubin, & Schouenborg, 2015).

These simulation pedagogies can be considered an implementation of Kolb's (1984) theory of experiential learning which suggests a cycle of four stages—experience, reflection, conceptualisation, experimentation. To enable effective learning it is not simply enough for a learner to have an experience in order to learn, but rather the learner must construct the links between theory and action; students create and then challenge their knowledge through reflection (Kolb & Kolb, 2005). The cycle can be entered at any stage but must be completed for learning to occur. The 'games' and simulations allow students to apply theories and concepts, experiment, and reflect on their learning. This experiential learning fosters the required emotional engagement in the subject for students to move beyond being passive observers to becoming active participants creating change for a more sustainable future (Dieleman & Huisingh, 2006). The exercises require students to work with others and develop their team working skills, and may help to draw in students who feel alienated by more 'traditional' models of teaching (Asal & Blake, 2006). Students are forced to role-play, requiring them to develop an understanding of their subject to make defensible arguments in support of their defined role. The simulations help develop their communication, negotiation, and research skills.

3   Case Study 1—'Design Your Own'

Fundamental to understanding standardisation is the knowledge of the process of standard creation and of standard function. This exercise was initially run as a second-year university exercise, for students studying on an environmental sciences curriculum. The exercise is run at a point where students have achieved a basic understanding of sustainability issues and serves to introduce the role of standardisation in sustainable development. It requires students to inhabit the roles and perspectives of different actors in the standardisation process. Students are challenged to recognise areas of their daily lives affected by standards—commonly recognising areas such as fast food recipes, mobile phone communications, and video format standards. This is developed to consider how standards might affect industries, and how indeed many standards are vital to the function of technology and services. Students then discuss what are the characteristics of these standards, how the standards are agreed and formed, and what conflicts might arise between different actors in the standards process. Following from this brief introduction student groups are challenged to develop a proposal for a standard to solve a sustainability issue of their choosing. The exercise is time constrained (a 1-h session works well). The issue can be industry specific or a wider issue relating to the SDGs. Examples proposed by students for standards include standards for modular construction processes, efficient manufacturing, and education standards.

To simulate the standards development process students are required to present their ideas as a new standard proposal to an 'expert' panel, comprised of teachers and fellow students. The panel question the proposal and provide feedback on the concept and whether it has the basis of a standard. This aspect of the exercise can be greatly enhanced when run with input from national standards bodies providing expertise and input from real-world case studies.

This exercise can be extended with a further group exercise to simulate the standards development process. With students playing the role of different actors in the development process (e.g. subject experts, industry lobbyists, standards authors, standards body). Each student or group is provided with a prepared perspective briefing specific to their role on the issue—for example a lobbyist or industry representative may be required to seek the 'watering-down' of the standard. The class then seeks to agree the basis for the 'new' standard through consensus or a more formal voting means. Student representatives are given a time-constrained slot to present their perspective, with a general forum after the presentations. The negotiation and voting stages can be facilitated through simple tallies, or through more complex means such as Twitter and other interactive technologies (Kemp et al., 2015). The challenges inherent in attempting to reach consensus on the role, reach, and function of standards are a core component of this exercise. There is a potentially significant opportunity for students to understand the complexity behind both standards development; the mixed—constraining in and enabling effect of standards for innovation; negotiation and consensus, and the deep challenges of sustainable development.

4 Case Study 2—Carbon Footprints

Carbon footprints have become a commonly used metric to relate a unit of human activity with an amount of GHG emission (Williams, Kemp, Coello, Turner, & Wright, 2012; Wright, Kemp, & Williams, 2011). The metric developed from the concept of 'ecological footprinting'—a conceptual representation of the area of land required to provide necessary resources and assimilate pollution for a given activity or population (Wackernagel & Rees, 1996). However in the context of the carbon footprint, the term 'footprint' does not apply to a literal area or amount of land, instead being representative of the Global Warming Potential ($kgCO_2e$) of the subject under study (Williams et al., 2012). Where GWP values, maintained by the Intergovernmental Panel on Climate Change (IPCC), express the equivalent amount of CO_2 for a given GHG that would have the same effect on radiative forcing if emitted to the atmosphere (IPCC, 2013). The rapid uptake of carbon footprints has prompted response in the standard community with standards emerging in companies and from industry collaborations (e.g. The Greenhouse Gas Protocol, 2005), as Publicly Available Specifications [e.g. PAS2050, PAS2070 (British Standard Institute, 2013); The British Standards Institution, 2011], and more recently as international standards (ISO/TS 14067:2013).

The proliferation of the carbon footprint has led to confusion regarding methods, applications, and importantly what the term actually means (Turner et al. 2012; Wright et al., 2011). This means that the use and application of the relevant standards is vital to common approaches and application. It also provides an excellent route for the exploration of the role of standardisation to ensure effective methodological approaches, calculation, and reporting.

Primarily, this exercise was run for university students in their first year of study; developing knowledge and understanding of carbon footprints and their underlying theory, as well as an appreciation of the role of standards. The exercise initially avoids reference to standards entirely, instead focusing on the underlying theory and methods of carbon footprints (see Williams et al., 2012 for example). Following a brief introduction, students are provided with a simplified 'real-world' carbon footprinting project, whereby they undertake an energy and waste audit of their university or college. They then apply given emission factors to calculate a carbon footprint. It is assumed that students have underlying knowledge in climate change and the role of GHG emissions.

Students are given freedom to design the energy and waste audit methodology and set boundaries as they deem necessary. Once completed the exercise is followed by a group discussion exercise to compare results. Commonly, students will return variations on the methods applied, and importantly the boundaries and results achieved.

This enables a discussion about the issues and failures created through not applying a common approach, and subsequently the need for standards. Key discussion points include the range of applicable standards, commonalities and differences (i.e. standard approaches) in student approaches and the applicable standards, the benefits and costs of a standardised approach.

5 Case Study 3—Life Cycle Assessment

Life cycle assessment (LCA) is a systematic approach to inventory and measure the environmental impacts of a product, process, or service. The term 'life-cycle' refers to the concept of producing a holistic assessment of all stages of a product, including: raw material abstraction and processing, manufacture, distribution, use, and disposal, and all intervening transportation steps (commonly referred to as 'cradle-to-grave'). The outputs of the assessment can be applied to identify areas of environmental impact relating to a product, to optimise environmental performance, or to compare two or more product options (Schenck & White, 2014).

LCA has a long history, being conceived in the late 1960's with similar ideas being developed in both the USA and Europe. The Coca-Cola Company provides one of the earliest examples being a study that attempted to quantify the energy, material, and environmental consequences of the entire life cycle of beverage packaging. The company was interested in whether they should self-manufacture of beverage cans among other issues relating to the packing process. The study was never published because of its confidential content, but was used by the company to inform their business decisions (Hunt, Franklin, & Hunt, 1996). While the development and application of holistic environmental methodologies to inform business decisions is a positive step towards a more sustainable world, standardization can play a role in ensuring that methodologies are robust and correctly applied. This case study forms the basis of discussions with students to highlight the potential consequences of an ad hoc approach to comparative studies and a failure to disclose results and information.

Concerns regarding the use of LCAs to make broad and often inappropriate marketing claims, pressure including legal censure led to a concerted effort led by the SETAC (Society of Environmental Toxicology and Chemistry) and the standardisation of LCA by ISO (International Standardisation Organisation) to develop a suitable LCA approach (Klöpffer, 2014). This action culminated in the development of the LCA standards in the ISO 14000 series (ISO 14040, ISO 14041, ISO 14042, ISO 14043). The standards were later revised and condensed into two standards—ISO14040:2006 Environmental management—Life cycle assessment—Principles and framework; and ISO14044:2006 Environmental management—Life cycle assessment—Requirements and guidelines.

The rich history of LCA and the ongoing development in the field serve to illustrate the complexities inherent in the method and the related standards. Importantly, in a classroom situation this quickly makes learning directly by the standard difficult and often inappropriate. Commonly classes, this case example included, make use of a case study or redesign example (Cooper & Fava, 1999). The standards then serve as the background rather than the forefront of the class. Students learn the underlying theory and practice of LCA with some contextual development of the related standards.

This class was developed for students studying on a graduate level course but has since been delivered as a pre-conference workshop, and for students at undergraduate levels. In this example, students are presented with several preliminary exercises in 'life-cycle' and 'systems' thinking—developing understanding of the supply and manufacturing chains of products. This is followed by a more detailed study of a packaging problem. As a group exercise, students map the life cycle of the packaging (in this case a basic milk carton). To simplify the exercise students are presented with a collection of pre-printed cards representing stages in the production life-cycle. They arrange the life-cycle, then discuss what should and should not be included in the study (e.g. should cattle farming be excluded from the system?). This then introduces them to first component part of the standards—goal and scope. The exercise then requires students to agree a functional unit for the study, map the emissions sources and sinks across the life cycle, and apply basic flow quantities provided by the instructor. The group is then guided through the process of calculating emissions relative to the functional unit—introducing them to the second stage of the standard—inventory analysis.

Students are then provided with a guided example of environmental impact calculation. They then work in groups to calculate values for a particular impact category (e.g. GWP) based on the data established earlier in the class and the impact factors provided. This then allows reflection against the third key component of the standard—impact assessment. The final stage requires students to interpret the results and provide context and recommendation based on their assessment—i.e. the fourth stage—interpretation. In this manner students are introduced to the process of LCA, in accordance with the standards, without initially directly referencing the standards per se, but with a clear and developed understanding of the importance and application of the standards.

6 Discussion and Conclusions

The three case studies presented offer methods to develop knowledge and understanding of sustainability standards. The primary aims of the case studies presented are to enhance understanding of sustainability tools and problems, and to develop understanding of standards and their role in sustainable development. Dealing with three broadly similar topics, the techniques have a common underlying theme of simulation and gaming as a core mechanism for the development of student understanding. Simulation and games in education represent an increasingly popular and important mode of learning. The simulations presented can be as complicated as desired to more closely approximate 'real-world' conditions or simplified to reflect desired learning. Simulations as part of a package of learning methods can reach

Standards in the Classroom: A Vehicle ...

students who might otherwise be unready or unwilling to follow a more abstract presentation. As one student stated following partaking in the 'Design your own' exercise:

> I like how the exercise isn't assessed ...I felt I could join in and make mistakes and experiment ...

The methods presented offer the ability to develop contextual and theoretical understanding of sustainable development tools, approaches and the role of standards. The majority of students will gain from the experience, moving beyond the traditional lecture format and enabling students to become active participants significantly increases retention of learnt information (McIntosh, 2001).

The simulation exercises, by definition, require some form of simplification—in the first case study students propose and negotiate a standard in a handful of hours; whereas the real-world development process can take several months, or longer. In the cases presented, the first serves to demonstrate the principle structures of standards. Including the need for standardisation and the role standards play in areas such as trade, technology, and sustainability. The emphasis of the exercise is primarily on the role of standards, and thus the 'committee' stage is broadly rule-free; but the exercise can be furthered to develop greater complexity in the negotiation process. More closely reflecting the conditions of national and international standardisation committees. Through this part of the exercise students develop an understanding and appreciation of different viewpoints, as a student stated following the 'design your own' exercise:

> The exercise encourages us to communicate and negotiate with each other to try to understand others views and reach an agreement. I really understand why there is such disagreement [about sustainability and standards] ...

The latter two cases deal more with the underlying theories, choosing to apply standards as conceptual structures for student learning. This helps develop a deep understanding of the standard as a framework to ensure comparability, interoperability, and compatibility of techniques, methods, and outcomes.

To ensure the success of the exercises clear instruction to students with an introduction to the topic or the scenario is vital. In the context of the 'design your own standard' this comprises a brief discussion regarding the role and purpose of standards. This provides the student with enough information to consider the purpose of a standard and offer their own thoughts on processes or systems they feel should be standardised. Interestingly, in this exercise students often inadvertently highlight proposals for standards that already exist. This can be used in the exercise to discuss the proliferation of standards in modern life.

These methods also have their shortcomings, and ideally should not be viewed as a method in isolation, but as part of a defined package of learning. Importantly, active methods of learning such as those presented require substantially more input from instructors and students, both in, and outside, the classroom. Simulation and gaming within education require precise planning for effective delivery. These exercises simplify complex processes and relationships where misrepresentation or misunderstandings may occur and lead to failure of the learning objectives. The role of simulations in the learning process should not dominate the intended learning process, else it risks students fixating on the game and missing the desired learning (McIntosh, 2001).

Indeed, as in the LCA case study presented it is often prudent to stop or divide the exercise into discreet sections to allow students to reflect. As participants of the exercise, run as a conference workshop, reflected:

> I liked being able to stop and start, and to discuss each stage of the process…the end session really helped clarify what we had done.

> The practical meant we were able to keep up effectively by discussing each stage and then demonstrate what we had understood

This can be particularly useful when considering standards as they commonly offer natural breaks in content. For example, the four-stage approach applied through ISO14040 (goal and scope, inventory analysis, impact assessment, interpretation) provides distinct points where simulations can be stopped/started to allow development of theoretical content. This contextualisation is critical to successful simulation exercises as they should seek to replicate reality as closely as possible, without unnecessarily encumbering the learning process with additional complexity (Dieleman & Huisingh, 2006).

These debriefings and discussions are vital to provide students with the opportunity to internalise and reflect on the lessons of the exercise. Without the opportunity to reflect on their learning students will not complete the learning cycle and may not make the connections between the concepts they have experienced during the exercise (Asal & Blake, 2006; Kolb, 1984).

The three case studies presented are based on simulation exercises to encourage knowledge and understanding of the underlying theoretical aspects of tools for sustainable development, and the role of standards. Significant value exists in these formative simulation pedagogies, which can help students understand how political and industrial negotiation and decision making in standards will have differing environmental, social, and economic consequences for sustainable development.

Whilst it is often stated that the 'best standards are the ones that go unseen', perhaps in the context of sustainable development the best standards are the ones that make an impact. In the field of climate change and environmental life cycle assessment, standards play a prominent role both in policy and business. It is vital therefore that students are equipped with the knowledge and skills to understand the complex issues of sustainable development, and the role of standards in helping address them.

Through the interactive style of simulation pedagogies students have the opportunity to put their learning into practice. Developing theoretical understanding in the context of the relevant standard and reflecting on the mechanisms that enable the standard to operate effectively. Moving student observers to active participants encourages learning, enhances employability, and positions them to effectively contribute to a more sustainable world through involvement in policy, education, and business applications.

References

Asal, V., & Blake, E. L. (2006). Creating simulations for political science education. *Journal of Political Science Education, 2*(1), 1–18.

Bíró, G. I. (2014). ScienceDirect didactics 2.0: A pedagogical analysis of gamification theory from a comparative perspective with a special view to the components of learning. *Procedia—Social and Behavioral Sciences, 141,* 148–151.

Brunsson, N., Rasche, A., & Seidl, D. (2012). The dynamics of standardization: Three perspectives on standards in organization studies. *Organization Studies.*

BSI (British Standard Institute). (2013). *PAS 2070:2013 specification for the assessment of greenhouse gas emissions of a city—Direct plus supply chain and consumption-based methodologies.*

Cooper, J. S., & Fava, J. (1999). Teaching life-cycle assessment at universities in North America. *Journal of Industrial Ecology, 3*(2/3), 13–17.

Dieleman, H., & Huisingh, D. (2006). Games by which to learn and teach about sustainable development: Exploring the relevance of games and experiential learning for sustainability. *Journal of Cleaner Production, 14*(9–11), 837–847.

Egyedi, T.M., & Ortt, J. R. (2017). Towards a functional classification of standards for innovation research. In R. Hawkins, K. Blind, & R. Page (Eds.), *Handbook of innovation and standards.* Edward Elgar.

Glover, D., Law, S., & Youngman, A. (2002). Graduateness and employability: Student perceptions of the personal outcomes of university education. *Research in Post-Compulsory Education, 7*(3), 293–306.

Glynn, S. M., Aultman, L. P., & Owens, A. M. (2005). Motivation to learn in general education programs. *The Journal of General Education, 54,* 150–170.

Horn, L., Rubin, O., & Schouenborg, L. (2015). Undead pedagogy: How a Zombie simulation can contribute to teaching international relations. *International Studies Perspectives, 17*(2), 187–201.

Hunt, R. G., Franklin, W. E., & Hunt, R. G. (1996). LCA-how it came about LCA—How it came about—Personal reflections on the origin and LCA in the USA. *International Journal of Life Cycle Assessment, 1*(1), 4–7.

IPCC. (2013). Climate change 2013: The physical science basis. In T. F. Stocker, D. Qin, G. -K. Plattner, M. Tignor, S. K. Allen, J. Boschung, A. Nauels, Y. Xia, V. Bex & P. M. Midgley (Eds.), *Contribution of working group I to the fifth assessment report of the intergovernmental panel on climate change.* Cambridge, United Kingdom and New York, NY, USA: Cambridge University Press

Kemp, S., Martin, F., & Maier, P. (2008). A gap analysis of student employability profiles, employer engagement and work-placements. *Planet, 21,* 16–20.

Kemp, S. (2009). Embedding employability and employer engagement into postgraduate teaching a case study from environmental management systems. *Planet, 21*(1), 47–52.

Kemp, S., Kendal, J., Warren, A., Wright, L., Canning, J., Grace, M., & Saunders, C. (2015). Global consensus is a dream, but twitter is real: Simulating a sustainable development goals summit through interdisciplinary classroom politics and negotiation by social media. In W. Leal Filho et al. (Eds.), *Integrative approaches to sustainable development at university level.* World Sustainability Series. Springer: Switzerland.

Klöpffer, W. (2014). Introducing life cycle assessment and its presentation in 'LCA Compendium'. *Background and future prospects in life cycle assessment* (pp. 1–37). Dordrecht: Springer.

Kolb, D. (1984). *Experiential learning: Experience as the source of learning and development.* Englewood Cliffs, New Jersey: Prentice-Hall.

Kolb, A. Y., & Kolb, D. A. (2005). Learning styles and learning spaces: Enhancing experiential learning in higher education. *Academy of Management Learning & Education, 4*(2), 193–212.

Komives, K., & Jackson, A. (2014). Voluntary standard systems. In C. Schmitz-Hoffmann, M. Schmidt, B. Hansmann, & D. Palekhov (Eds.), *Voluntary standard systems.* Berlin: Springer.

Kumar, R., & Lightner, R. (2007). Games as an interactive classroom technique: Perceptions of corporate trainers, college instructors and students. *International Journal of Teaching and Learning in Higher Education, 19*(1), 53–63.

Manning, S., Boons, F., Von Hagen, O., & Reinecke, J. (2012). National Contexts Matter: The Co-Evolution of Sustainability Standards in Global Value Chains. *Ecological Economics, 83,* 197–209.

McIntosh, D. (2001). The uses and limits of the model United Nations in an international relations classroom. *International Studies Perspectives, 2*(3), 269–280.

Prince, M. (2004). Does active learning work? A review of the research. *Journal of Engineering Education, 93*(3), 223–231.

Reinecke, J., Manning, S., & Von Hagen, O. (2012). The emergence of a standards market: Multiplicity of sustainability standards in the global coffee industry. *Management and Marketing, 33*(56), 789–812.

Rittel, H. W. J., & Webber, M. M. (1973). Dilemmas in a general theory of planning. *Policy Sciences, 4,* 155–169.

Schenck, R., & White, P. (2014) Environmental life cycle assessment measuring the environmental performance of products. In R. Schenck & P. White (Eds.), *American center for life cycle assessment.*

Stenzel, P. L. (2000). Can the ISO 14000 series environmental management standards provide a viable alternative to government regulation? *American Business Law Journal, 37*(2), 237–298. https://doi.org/10.1111/j.1744-1714.2000.tb00272.x

The British Standards Institution. (2011). *Publicly available specification PAS 2050: 2011 specification for the assessment of the life cycle greenhouse gas emissions of goods and services.*

Turner, D., Williams, I., Kemp, S., Wright, L., Coello, J., & McMurtry, E. (2012). Towards standardization in GHG quantification and reporting. *Carbon Management, 3*(3), 223–225.

UNA-UK. (2016). *Model UN Portal | UNA-UK.* Available at: https://www.una.org.uk/get-involved/learn-and-teach/model-un-portal. Accessed December 22, 2017.

United Nations General Assembly. (2015). *Transforming our world: The 2030 agenda for sustainable development.* Available at: https://sustainabledevelopment.un.org/content/documents/7891Transforming%20Our%20World.Pdf. Accessed December 22, 2017.

Wackernagel, M., & Rees, W. E. (1996). *Our ecological footprint : Reducing human impact on the earth.* New Society Publishers.

Williams, I., Kemp, S., Coello, J., Turner, D. A., & Wright, L. A. (2012). A beginner's guide to carbon footprinting. *Carbon Management, 3*(1), 55–67.

World Resources Institute and World Business Council for Sustainable Development, & The Greenhouse Gas Protocol. (2005). *The GHG protocol for project accounting.* Washington, USA: World Resources Institute.

Wright, L. A., Kemp, S., & Williams, I. (2011). "Carbon footprinting": Towards a universally accepted definition. *Carbon Management, 2*(1), 61–72.

Laurence A. Wright is a Senior Lecturer in the Warsash School of Maritime Science and Engineering at Solent University. His work is in understanding and addressing environmental impacts from diverse human activities and products; from beer production to the role of the maritime industry in tackling climate change. His expertise is centred on the concepts of life cycle assessment, circular economics, resource efficiency and sustainable development. Laurie obtained a BSc

Environmental Sciences, an MSc Environmental Monitoring and Assessment, and a PhD examining life-cycle community GHG emissions from the University of Southampton. He is a Fellow of the Higher Education Association (FHEA), and a member of the American Center for Life Assessment (ACLCA) and co-chair of the society's Education Committee. He was awarded the ACLCA award for Academic Leadership 2016.

Julie Sinistore is a project director, on the Sustainability, Energy and Climate Change team at WSP USA and has been conducting life cycle assessments (LCAs) since 2007. Her expertise includes LCA and carbon and water Footprinting of products and services, in addition to sustainability strategy in the electronics, bio-based fuels and chemicals, agriculture, and building and construction sectors. She holds a Ph.D. in Biological Systems Engineering and a M.Sc. in Agroecology from University of Wisconsin-Madison, and a B.S. in Natural Resource Management from Cook College of Rutgers University (both land grant schools). Her research work and publications focus on sustainable agricultural practices, LCA of bio-based fuels, energy policy analysis, and sustainability assessment methodology. Julie has taught courses on various topics, including Life Cycle Thinking, Sustainable Product Design, Renewable Energy, and Dendrology at UC Berkeley, UW Madison, and Rutgers University. She is a member of the board of the American Center for Life Cycle Assessment and serves as vice chair of the Education committee. She is also a certified individual Environmental Product Declaration (EPD) verifier with The International EPD® system. Julie also serves on the ISO/TC Technical Advisory Group (TAG) 207 on Environmental Management which provides guidance on revisions to the ISO standards related to LCA.

Mike Levy is the senior associate for First Environment, an environmental engineering and consulting firm providing cutting edge solutions since 1987. He has over 40 years experience in the sustainability and life cycle field, working as an LCA consultant, for private industry, and managing LCA programs for a major trade association representing the chemicals and plastics industry. Prior to First Environment, he was the senior director for a group within the American Chemistry Council (ACC)—the Plastics Foodservice Packaging Group (PFPG), a national organization representing major producers and raw material suppliers of plastics food service and packaging products. Mike also serves as senior director, life cycle issues, for the Plastics Division of the ACC. Prior to joining ACC, Mike managed the Polystyrene Packaging Council (PSPC) as a program of the Society of the Plastics Industry, Inc. (SPI). Mike also served as director, energy and materials policy for the American Forest and Paper Association (AF and PA), representing the U.S. forest, wood, and paper industries. He managed energy, life cycle assessment, standards, and municipal solid waste policy issues for AF and PA. Mike holds undergraduate and graduate engineering degrees from Rensselaer Polytechnic Institute (RPI), and pursued an M.B.A. degree at Fairleigh Dickinson University. He is listed in Who's Who in America, Who's Who in Plastics and Polymers, and in the International Directory of Experts and Expertise. Mr. Levy is involved in several Life Cycle Assessment (LCA) programs: serving as an international expert to the ISO 14000 standards development (International Standards Organisation) on the Life Cycle Assessment activities; serving as the policy committee chair of the American Center for Life Cycle Assessment (ACLCA) and received its outstanding individual service award in 2012, and is a platinum member representing the Plastics Division of ACC in the UNEP/SETAC Life Cycle Initiative program. He is a Certified Lifecycle Executive (CLE) under the ACLCA. He was recently elected as one of the co-chairs for the US TAG for the ISO TC 323 Circular Economy standards activity for the U.S. delegation. He has written numerous papers and publications on environmental issues and serves on the advisory boards of several trade publications.

Bill Flanagan is Co-founder and Director of Aspire Sustainability, a consultancy focusing on strategic sustainability solutions including product life cycle assessment. He was previously Founder and Director of GE's Eco-assessment Center of Excellence and was responsible for developing GE's product sustainability strategies from 2007 to 2017. Bill has extensive leadership and teaming experience and has conducted LCA studies in a wide range of product categories including renewable energy (wind, solar, hydro), aircraft engines, biofuels, LED lighting, consumer appliances, batteries, locomotives, advanced materials, additive manufacturing, CNG and LNG technologies, biopharmaceutical processing, gas and steam turbines and generators, anesthesia gases and equipment, remanufacturing, medical equipment (CT, MR, digital X-ray, ultrasound), subsea oil and gas technologies, and packaging materials. In 2014 Bill was awarded the Lifetime Individual LCA Leadership Award by the American Center for Life Cycle Assessment (ACLCA). In the same year, Bill was elected to serve as Chair, Board of Directors, ACLCA, where he has fostered a community of renewed excitement and collaboration. Bill graduated from Virginia Tech and received a Ph.D. in Chemical Engineering from the University of Connecticut. He is a certified life cycle assessment professional (LCA CP).

A Look Behind the Curtain: Standardization Education for Engineers from the Electric Vehicle Standardization Shopfloor

Peter Van den Bossche

1 Engineering Education in Brussels: The Bruface Programme

The Brussels Faculty of Engineering (*Bruface*) is an initiative of the two main universities located in Brussels, Belgium. The French-language *Université Libre de Bruxelles* (ULB) was founded in 1834 and saw its Dutch-language section, the *Vrije Universiteit Brussel* (VUB), becoming independent in 1969. Since 2010 however, VUB and ULB strengthened their collaboration over community borders and jointly offer a broad spectrum English-language master programme in engineering called Bruface (Bruface, 2018). Building a challenging curriculum for future engineers in a changing technological landscape is one of the missions of Bruface, enabling the student to build a broad scientific knowledge that combines a multidisciplinary engineering training with an in-depth specialisation in the chosen major. The programme trains engineers in scientific and technological efficiency. The programme is academic, meaning that it is characterized by close links to scientific research in the related fields as well as the profession. Students must obtain a scientific balance between thorough, critical knowledge and practical skills, with emphasis on independence, creativity and inventiveness.

One of the majors offered is *Vehicle Technology and Transport*, forming engineers who can design systems in which transportation of people and goods are central, with special attention to innovative, environmentally friendly electrically propelled vehicles. This major builds on VUB's long tradition in electric vehicle research, starting in the 1970s under the initiative of the late Professor Gaston Maggetto. In urban traffic, due to their beneficial effect on environment, electrically propelled vehicles are an important factor for improvement of traffic and more particularly for a healthier living environment. VUB's MOBI research group has now developed into

---

P. Van den Bossche (✉)
Vrije Universiteit Brussel, Pleinlaan 2, 1050 Brussel, Belgium
e-mail: pvdbos@vub.ac.be

© Springer Nature Switzerland AG 2020
S. O. Idowu et al. (eds.), *Sustainable Development*, CSR, Sustainability,
Ethics & Governance, https://doi.org/10.1007/978-3-030-28715-3_2

the premier Belgian research centre on sustainable transport, focusing on energy storage, power electronics, environmental impact, transport logistics, and also standardization, where since 25 years VUB experts participate in international technical committees concerning electrically propelled vehicles and their infrastructure, with VUB now holding the secretariat of technical committees IEC TC69 and CENELEC TC69X.

2 Course on Standardization

2.1 Introduction and Selected Approach

Standardization has tradionally known a limited coverage in the engineering curriculum. With the need for standardization education being internationally recognized and discussed [cf. (Hesser & de Vries 2011; Olshefsky, 2008)], steps were taken to introduce it into the VUB curriculum.

A first introduction to standardization for VUB engineering students is given in the framework of the electrotechnical course on electrical installations in the bachelor programme. A deeper exploration of the subject was developed for the master programme, through the establishment of an elective course (3 ECTS).

The main focus and highlights of this course had to be defined. There are many examples to be found of standardization course material proposed (UNECE, 2012) or already developed (APEC SCSC, 2008), these were however for a large part aiming at business and economy-oriented studies focusing on the macro-level and studying among others the economical and societal aspects of standardization, or part of more extensive programmes going beyond the scope of a 3 ECTS course (APEC SCSC, 2011; UNECE, 2014).

To make the elective course appealing to engineering students, a more technical approach of standardization has been selected, taking a bottom-up view focusing on actual standards development. This allowed to take maximal advantage of the available expertise about the standardization process and more accurately addressed student's expectations and interests.

The course can be divided in three parts:

- General introduction about standardization;
- Specific study of electric vehicle standardization, as an interesting example of standardization in a developing technology field.
- Practical exercises where students can get a hands-on experience of the standards drafting process.

The first emanation of the elective course started from 2008, as *Reglementering en normalisatie* (Regulation and standardization) presented in the Dutch-language program; with the introduction of the Bruface Vehicle Technology major this course is now taught in English as *Automotive standardization*, consisting of both lectures and practical exercises. Students enrolling in this course typically hold a bachelor in Engineering Science or Engineering Technology, most from the VUB and ULB although there is a growing influx of foreign students with different but equivalent backgrounds.

2.2 Lectures on General Standardization Issues

The general introduction about standardization is inspired by the effort of NBN, the Belgian National Standards Body, to promote standardization in education (NBN, 2018), a typical example of the collaboration established between NSBs and universities promoted by international standards bodies (ISO, 2014). In this framework, specific introductory course materials (to which the author has contributed) have been developed.

For the master course *Automotive standardization*, a thorough introduction is given, informing the master students on the main characteristics of the subject. The introductory lectures highlight main topics which can be summarized as follows:

- *What are standards*—there are many misconceptions about this issue on all levels. Press articles and even statements by politicians often use the term *standard* in an improper way, for example mentioning concepts like *the standards of the European Commission*, whileas this institution does not publish any standards proper (although a number of regulatory and communication documents may in practice function as standards). The official definition of a standard (up for discussion but still valid, and a good reference specifically for formal standards) as *a document, established by consensus and approved by a recognized body, that provides, for common and repeated use, rules, guidelines or characteristics for activities or their results, aimed at the achievement of the optimum degree of order in a given context* (ISO/IEC Guide 2, 2004), offers the necessary guideline in this matter.
- *What kind of standards are there around*—with product, process, management and other types. For the standards which are relevant for this course, the main subjects encompass one or more of the areas of safety, compatibility and performance, as illustrated in the *house of standardization* [Fig. 1; (Van den Bossche, 2010)]. One may imagine more pillars to this house (such as management standards), but these are the main ones considered in this course.
- *Who is making standards*—highlighting the global, European and national standardization bodies, their structure, operation and interaction. The course focuses on the *committee-based standards* (Wiegmann, de Vries, & Blind, 2017), particularly those from the IEC, as these are the most relevant ones in the field. Market-based standards (such as for example the CHAdeMO DC charging system) are present but typically they become eventually absorbed in the committee standards [in case (IEC61851-23, 2014)], whileas government action is mainly related to the implementation of the international standards as European standards in the framework of the European directives—an action which, for most electrotechnical standards,

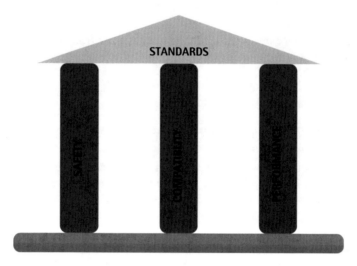

Fig. 1 The house of standardization

typically takes place after achievement and publication of the international standard.

- *How standards come to being*—explaining the bottom-up, consensus based process which is one of the major strengths of standardization as it allows to create broadly accepted and truly representative standard documents. However, the weakness of this process is the slow nature of standards drafting. The drafting process of standards is further explored in the practical exercises.
- The highlighting of this aspect of standards development procedures and practices is often not present in engineering-oriented courses on standards, who tend to focus on purely technical perspectives (ISO, 2014). The underlying course takes an unique approach in this field.
- *What is the difference between standards and regulations*—there are many misconceptions on this point, with laws and regulations being enforceable and standards—in principle—voluntary, however there are many interactions between the two realms. Specific cases include on one hand the European *New Approach* (Pelkmans, 1987), with EU directives stating essential requirements and harmonized standards presuming conformity. This process particularly affects the adoption of international (IEC) standards as European (CENELEC) ones, with, at least in the electrotechnical field, most of the work being done on a global (IEC) level with parallel voting in CENELEC. On the other hand, there is the specific situation concerning road vehicles which are not subject to New Approach directives, but where UNECE regulations (UNECE, 2018) are used for type approval.
- *What is the relevance of standards for society and economy*—the impact of standardization goes well beyond the mere definition of product characteristics and can be particularly relevant when considering emerging technologies and innovation.

Although the focus of the course, being within the engineering curriculum, is mainly on the technical aspects of standards, and particularly on the drafting process, and thus less on the economical and societal aspects, this is an essential part of the introduction.

2.3 The Electrically Propelled Vehicle and Its Standardization

In subsequent lectures, students learn about the standardization of the electrically propelled vehicle. Electric drives, energy storage and charging infrastructure represent some of the major research interests of VUB which is reflected in the research-driven master programme in transportation technology.

The focus on this subject for standardization studies offers some interesting opportunities. Through VUB's involvement in standardization work in the field, a direct feedback from the standards committees is possible which opens specific possibilities for the practical exercises. Furthermore, the subject itself is a very interesting one from a standardization point of view. Electric vehicles represent a technology coming to maturity and gaining a particular societal interest, with a lot of standardization work being developed on a global level.

The standards landscape for electrically propelled vehicles is a very complex one as it is multidisciplinary and involves various committees [Fig. 2, (Van den Bossche, 2016)]. A main issue is the question which standardization body should have the main responsibility for electric vehicle standards, the *International Electrotechnical*

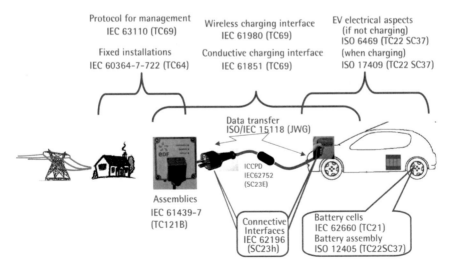

Fig. 2 EV standardization landscape

Commission (IEC) dealing with all things electrical, or the *International Organization for Standardization* (ISO) dealing with general technologies. This problem is less straightforward than it looks: the electric vehicle represents in fact a mixed technology, being both a *road vehicle* and an *electrical device*, each of both realms having, for historical reasons, a different *standardization culture*, and the different background of experts seating in the respective committees (Van den Bossche, 2003).

A generally clear division of labour has been achieved however, with ISO (TC22 SC21, later SC37) covering the vehicle-related aspects and IEC (TC69) covering electrical aspects including charging infrastructure. IEC TC69 is thus the main committee dealing with charging infrastructure standards. Specific aspects such as batteries and accessories are covered by the respective technical committees with whom liaisons are established.

This infrastructure standardization has known a strong development in recent years, centering on the IEC61851 family of standards. IEC61851-1 (IEC61851-1, 2017), the third edition of which was issued early 2017, with the maintenance procedure already started, gives the general requirements that serve as a basis for all the subsequent standards in the series. encompassing mechanical, electrical, communications, EMC and performance requirements for EV supply equipment conductively connected and used to charge electric vehicles. The aspects covered in this standard include the characteristics and operating conditions of the EV supply equipment, the specification of the connection between the EV supply equipment and the EV, the requirements for electrical safety for the EV supply equipment, the connection to fixed installations and requirements for basic communication for safety and process matters (Van den Bossche, Turcksin, Omar, & Van Mierlo, 2017).

One main concept introduced by IEC61851-1 are the so-called *charging modes*, defining the way of connecting the electric vehicle to the supply network. For AC charging, *Mode 3* with its control pilot protection is now universally used.

Accessories for charging are described by standards from the relevant committee IEC SC23H. The European directive 2014/94 on the deployment of alternative fuels infrastructure (European Union, 2014) has prescribed the use of Type 2 accessories (IEC62196-2, 2016) for AC charging points in Europe, with charging points fitted with a Type 2 socket outlet, accessible for vehicles fitted with a Type 1 inlet using a suitable charging cable. Standardization for AC charging has thus been well achieved Europe-wide. However, this was the result of a tedious process characterized by an interaction between different actors involved (the European standardization body CENELEC and the European Commission), as well as between distinct standardization mechanisms (market-based, non-market based and public) (Wiegmann, 2013).

A greater variety exists for DC charging however, where several families of accessories are described in the standard (IEC62196-3, 2014), among which the CHAdeMO and CCS types. Note that DC charging always uses fixed cables attached to the charging station. The accessories are thus connectors and vehicle inlets. In Europe, although the directive 2014/94 prescribes the CCS system, fast charging stations are mostly equipped with both CHAdeMO and CCS connections, often also with a high power AC connection (43 kW over Type 2 connector), due to the various types of vehicles present on the market.

One issue that remains to be solved is the common use of pins in a Type 2 inlet for both AC and DC, this allows to use the compact Type 2 inlet for both AC standard charging and DC fast charging. This is still under discussion and not yet defined in the standards, the concept being hard to accept for the electricity sector where this practice is quite new and its behaviour under fault conditions is considered, fearing the presence of DC fault currents, that the AC protective devices in fixed installations are not designed for, in the AC network. However, this system has already been successfully implemented by Tesla in Europe.

There is a demand for really high power charging (hundreds of kW) exceeding the envelope considered by 61851-23, not only for buses but also to allow ultra-fast charging of cars. This issue will involve several other committees (e.g. IEC TC20 for cables). The impact of such high-power devices on the grid shall also be taken into account.

The issue of bidirectional power transfer (e.g. vehicle-to-grid) is also under consideration and has not been addressed yet.

Standardization of batteries mainly focuses on two aspects: safety and performance tests. In each case, IEC is dealing with individual battery cells and modules (IEC62660-1, 2010; IEC62660-2, 2018; IEC62660-3, 2016; IEC/TR62660-4, 2016) whileas ISO focuses on the battery system as a component in the vehicle (ISO12405-3, 2014; ISO12405-4, 2018). This is an example of a good collaboration between IEC and ISO leading to complementary and not conflicting standards (Van den Bossche et al., 2009).

Measuring the performance of an EV traction battery necessitates the definition of proper load cycles corresponding to the actual use of the battery, which can be energy-oriented as in a battery-electric vehicle or power-oriented as in a hybrid electric vehicle.

2.4 Practical Exercises

2.4.1 The Concept

To make the students really acquainted with the reality of standardization, and to offer them a more interesting experience rather than a dull approach, a specific concept has been designed which is quite unique as it gives students an insider's look into the innards of the standardization mechanism.

Selected recent standards related to electromobility are hereby analysed as a case study, where every student treats one particular example and has to describe the genesis of the standard.

The documents made available to be analyzed for each project typically include working documents illustrating the whole of the stages of standard genesis, as shown in Fig. 3:

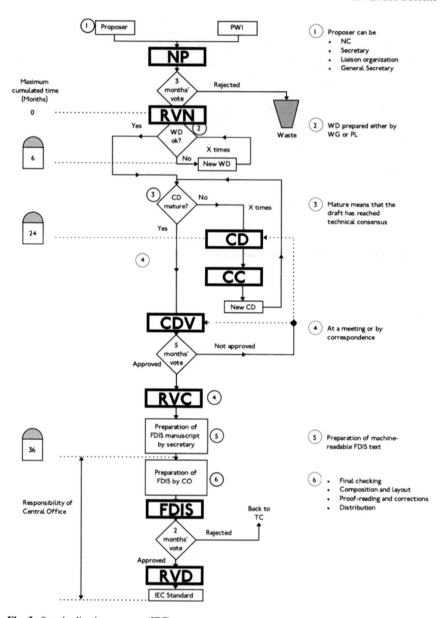

Fig. 3 Standardization process (IEC)

- the new work item proposal (NP)
- the result of voting on NP (RVN)
- the first committee draft (CD)
- the comments on this draft (CC)
- subsequent CD and CC (optional)
- the committee draft for voting (CDV)
- the comments on this draft (RVC)
- the final draft international standard (FDIS)
- the comments on the FDIS (RVD).

Access to these documents is made possible through the lecturer's activity in standardization committees. The documents are normally confined to project teams and standards committees, being only accessible to the *inner circle* of standardization. Students are well aware of this privileged access, understanding that draft standards are not to be disseminated to the general public, on one hand because they are liable to considerable changes which are reflected in the subsequent commenting rounds until consensus is reached, and on the other hand because they are copyrighted documents, standard sales being a key source of income for the standards bodies. Through accessing these documents the students thus get valuable insights in the processes which are not available to typical end users of the standard or to consumers.

By following the growth of the document from early proposal to mature standard, by analysing its evolution and in particular the comments expressed by national committees on the subsequent drafts (which in many cases reflect underlying economical or political motives), as well as the response by the project team processing such comments, background insights can be gained on the process and on the dynamic in which the standard is evolving, highlighting the roles of different actors in the process.

2.4.2 Examples

Selected projects cover a broad subject envelope pertaining EV standardization, as to allow students to choose a topic of their interest. They are chosen from recently published standards where the full set of project files (NP to RVD) is available. Some projects however are not withheld, on one hand due to their sheer complexity (such as IEC61851-1, 2017), a very complex standard with several CDs and thousands of comments) or overt simplicity (such as IEC62576, 2018), a standard drafted in a fast process without many technical comments).

Typical examples of the case studies include:

- IEC61851-21-1, 2017, IEC61851-21-2, 2018, dealing with EMC aspects of charging equipment
- IEC62196-1, 2014, IEC62196-3, 2014, SC23H documents dealing with charging accessories
- IEC62660-3, 2016, dealing with battery safety
- IEC62752, 2016, a SC23E standard on Mode 2 charging cables

- IEC62840-2, 2016, dealing with battery swap systems
- IEC62893-1, 2017, IEC62893-2, 2017, IEC62893-3, 2017, the TC20 standards about battery charging cables
- ISO17409, 2015, an ISO document, available for parallel voting in IEC TC69, offering insights in a vehicle-centric approach of EV standardization by ISO.

2.4.3 Evaluation

Rather than by a traditional exam, students are evaluated through the case studies performed, with both a written report on their assignment and a presentation to their peers (both on half of the points). This will report on the following:

- An analysis of the genesis process of the standard, covering subsequent versions and focusing on the origin of the comments and their resolution by the project team
- A short summary of the standard and its implications, based on the final draft version.

The report has to be strong and concise, focusing on essential aspects and selecting the most relevant items (rather than purely editorial comments for example), to be delivered in the accompanying presentation.

3 Conclusions

Whileas interest in this elective course was rather limited in the first few years (possibly due to the original title *Regulation and standardization* perceived as not sounding very attractive ...), it has since been growing, even appealing to students from other majors like *Industrial Sciences*.

The approach of this course and its practical exercises is unconventional, going beyond a classical textbook with exam. Whileas the official learning outcomes of the course are to offer students an overview of the world of standards and regulations and to make them capable of identifying and interpreting specifically applicable standards of a technical problem, the course allows them to understand the real meaning and impact of standards and to get real-life insights in a world which is largely unknown to them, through direct feedback from the standardization shopfloor. This gives students a considerable added value contributing to their skills and competences.

It is hoped that this course may incite students to consider participating in standardization work in their professional life, and thus contribute to the many benefits standardization has to offer for technology and society.

References

APEC SCSC. (2008). *Education Guideline 1: Case studies on how to plan and implement standards education programs and strategic curriculum model.* Asia-Pacific Economic Collaboration.

APEC SCSC. (2011). *Education Guideline 4: Teaching standardization in Universities: Lessons learned from trial program.* Asia-Pacific Economic Collaboration.

Bruface. (2018). *Brussels faculty of engineering.* Retrieved 2018-05-15, from http://www.bruface.eu.

European Union. (2014). *Directive 2014/94/EU of the European Parliament and of the Council on the deployment of alternative fuels infrastructure* (Vol. 57) (No. L307). OJ L307, 2014-10-28.

Hesser, W., & de Vries, H. J. (2011). *White paper on academic standardisation education in Europe* (EURAS, Ed.). EURAS.

IEC61851-1. (2017). *Electric vehicle conductive charging system–Part 1: General requirements* (3.0 ed.) [International Standard]. IEC.

IEC61851-21-1. (2017). *Electric vehicle conductive charging systems—Part 21-1: Electric vehicle onboard charger EMC requirements for conductive connection to AC/DC supply* (1.0 ed.) [International Standard]. IEC.

IEC61851-21-2. (2018). *Electric vehicle conductive charging system—Part 21-2: Electric vehicle requirements for conductive connection to an AC/DC supply—EMC requirements for off board electric vehicle charging systems* (1 ed.) (No. 69/531/FDIS). IEC TC69.

IEC61851-23. (2014). *Electric vehicle conductive charging system—Part 23: d.c. electric vehicle charging station* (1 ed.) [International Standard]. IEC.

IEC62196-1. (2014). *Plugs, socket-outlet and vehicle couplers—Conductive charging of electric vehicles—Part 1: Charging of electric vehicles up to 250 A a.c. and 400 A d.c.* (3.0 ed.). IEC.

IEC62196-2. (2016). *Plugs, socket-outlet and vehicle couplers—Conductive of electric charging c vehicles—Part 2: Dimensional interchangeability requirements for pin and contact-tube accessories with rated operating voltage up to 250 V a.c. single phase and rated current up to 32A* (2 ed.) [International Standard]. IEC.

IEC62196-3. (2014). *Plugs, socket-outlet and vehicle couplers—Conductive charging of electric vehicles—Part 3: Dimensional interchangeability requirements for d.c. and a.c./d.c. pin and tube-type vehicle couplers* (1 ed.) [International Standard]. IEC.

IEC62576. (2018). *Electric double-layer capacitors for use in hybrid electric vehicles—Test methods for electrical characteristics* (2.0 ed.) [International Standard]. IEC.

IEC62660-1. (2010). *Secondary batteries for the propulsion of electric road vehicles—Part 1: Performance testing for lithium-ion cells* (1 ed.) [International Standard]. IEC.

IEC62660-2. (2018). *Secondary lithium-ion cells for the propulsion of electric road vehicles—Part 2: Reliability and abuse testing* (1 ed.) [International Standard]. IEC.

IEC62660-3. (2016). *Secondary lithium-ion cells for the propulsion of electric road vehicles—Part 3: Safety requirements of cells and modules* (1 ed.) [International Standard]. IEC.

IEC/TR62660-4. (2016). *Secondary lithium-ion cells for the propulsion of electric road vehicles—Part 4: Candidate alternative test methods for the internal short circuit test of iec 62660-3* (1 ed.) [Technical Report]. IEC.

IEC62752. (2016). *In-cable control and protection device for Mode 2 charging of electric road vehicles (IC-CPD)* (1 ed.) [International Standard]. IEC.

IEC62840-2. (2016). *Electric vehicle battery swap system—Part 2: Safety requirements* (1.0 ed.) [International Standard]. IEC.

IEC62893-1. (2017). *Charging cables for electric vehicles for rated voltages up to and including 0.6/1 kV—Part 1: General requirements* (1.0 ed.) [International Standard]. IEC.

IEC62893-2. (2017). *Charging cables for electric vehicles for rated voltages up to and including 0.6/1 kV—Part 2: Test methods* (1.0 ed.) [International Standard]. IEC.

IEC62893-3. (2017). *Charging cables for electric vehicles for rated voltages up to and including 0.6/1 kV—Part 3: Cables for AC charging according to modes 1, 2 and 3 of IEC 61851-1 of rated voltages up to and including 450/750 V* (1.0 ed.) [International Standard]. IEC.

ISO. (2014). *Teaching Standards—Good practices for collaboration between National Standards Bodies and universities.* Retrieved 2018-05-15, from https://www.iso.org/files/live/sites/isoorg/files/archive/pdf/en/teachingstandardsen-lr.pdf.

ISO12405-3. (2014). *Electrically propelled road vehicles—Test specification for lithium-ion traction battery packs and systems—Part 3: Safety performance requirements* (1 ed.) [International Standard]. ISO.

ISO12405-4. (2018). *Electrically propelled road vehicles—Test specification for lithium-ion traction battery packs and systems—Part 4: Performance testing* (1 ed.) [International Standard]. ISO.

ISO17409. (2015). *Electrically propelled road vehicles—Connection to an external electric power supply—Safety specifications* (1 ed.) [International Standard]. ISO.

ISO/IEC Guide 2. (2004). *Standardization and related activities—General vocabulary* [ISO/IEC Guide]. Geneva.

NBN. (2018). *Education about Standardisation.* Retrieved 2018-05-18, from https://www.nbn.be/en/education.

Olshefsky, J. (2008). The role of standards education in engineering curricula. In ASEE (Ed.), *American Society of Engineering Education's Fall 2008 Conference, Mid-Atlantic Section.* ASEE.

Pelkmans, J. (1987). The new approach to technical harmonization and standardization. *JCMS: Journal of Common Market Studies, 25*(3), 249–269. https://doi.org/10.1111/j.1468-5965.1987.tb00294.x.

UNECE. (2012). *Document ECE/TRADE/C/WP.6/2012/6.—Concept note for the workshop on "Introducing standards related issues in educational curricula", including a proposed model programme on standardization.* Retrieved 2018-05-15, from http://www.unece.org/fileadmin/DAM/trade/wp6/documents/2012/wp6201206E.pdf.

UNECE. (2014, 11). *Results on introducing standardization discipline, MSRU.* Retrieved 2018-05-15, from http://www.unece.org/fileadmin/DAM/trade/wp6/AreasOfWork/EducationOnStandardization/MGOU-introduction_of_standardization_discipline_results_Nov2014.pdf.

UNECE. (2018). *Vehicle regulations.* Retrieved 2018-05-15, from http://www.unece.org/trans/main/welcwp29.html.

Van den Bossche, P. (2010). Matching accessories: Standardization developments in electric vehicle infrastructure. In *EVS-25 Proceedings.* Shenzen.

Van den Bossche, P. (2003). *The electric vehicle: Raising the standards* (Unpublished doctoral dissertation). Vrije Universiteit Brussel.

Van den Bossche, P. (2016). Developments and challenges for EV charging infrastructure standardization. *World Electric Vehicle Journal, 8*(2), 557–563.

Van den Bossche P., Turcksin, T., Omar, N., & Van Mierlo, J. (2017, 10). International standardization of charging infrastructure: Achievements and new developments. In *EVS30 Proceedings.* Stuttgart.

Van den Bossche, P., Verbrugge, B., Van Mulders, F., Omar, N., Culcu, H., & Van Mierlo, J. (2009). The cell versus the system: Standardization challenges for electricity storage devices. *World Electric Vehicle Journal, 3*(1), 1–6.

Wiegmann, P. (2013). Combining different modes of standard setting—Analysing strategies and the case of charging electric vehicles in Europe. In EURAS (Ed.), *Euras Proceedings* (pp. 397–411). Brussels.

Wiegmann, P. M., de Vries, H. J., & Blind, K. (2017). Multi-mode standardisation: A critical review and a research agenda. *Research Policy, 46*(8), 1370–1386.

Peter Van den Bossche graduated as "civil mechanical-electrotechnical engineer" from the Vrije Universiteit Brussel, promoted in Engineering Sciences at the VUB on the Ph.D. thesis The Electric vehicle, raising the standards. Since 25 years he is active in the field of international standardization concerning electric vehicles and their infrastructure, participating to several technical committees in IEC and ISO, and currently performs the role of Secretary of IEC TC69 and CEN-ELEC TC69X. He is lecturer at the Vrije Universiteit Brussel's engineering faculty, in charge of bachelor and master courses on electrical engineering and standardization. His research interests include standardization, electric vehicle infrastructure and energy storage.

From Boring to Intriguing: Personal Perspective on Making Education About Standardization Appealing to Students

Vladislav V. Fomin

1 Background of the Author

My academic affiliation is the mainstream Management Information Systems (MIS). Within the broad field of MIS, I was among the first scholars at the change of the millennia to explore the role of standards and the process of making standards. My Ph.D. thesis defended in 2001 was titled "The Process of Standard Making. The Case of Cellular Mobile Telephony", and the paper "Standardization: Bridging the Gap Between Economic and Social Theory" in year 2000 (Fomin & Keil, 2000)—one of the first papers on standardization published in the proceedings of International Conference on Information Systems—is still being cited. For almost two decades now, the topic of standardization remained prominent in my academic work, with over 50 publications directly pertaining to this body of knowledge. Through the long-term involvement in venues organized by European Academy for Standardization (EURAS), publishing on topics related to standardization, inevitably and naturally there was a knowledge spillover to my teaching.

My active teaching career started in 2008, when I joined Vytautas Magnus University in Lithuania. Having teaching responsibilities at two different departments—the Faculty of Informatics and the Faculty of Economics and Management—I introduced topics related to standards into a number of courses I taught. Very quickly I discovered that students would consider the "standard" body of knowledge on standards, like the one issued by International Electrotechnical Commission (IEC) (Egyedi, 2007) and featuring classification of standards, information on standardization bodies, etc.—too dry or abstract. Put it simple, students considered the topic of standards to be boring. This students' perception had two direct negative implications for my teaching. First, students did not pay attention to what was taught to them. Second, seeing that students do not pay attention was a challenge for me as a lecturer—both

---

V. V. Fomin (✉)
Vilnius University, Kaunas, Lithuania
e-mail: vvfomin@gmail.com

© Springer Nature Switzerland AG 2020
S. O. Idowu et al. (eds.), *Sustainable Development*, CSR, Sustainability,
Ethics & Governance, https://doi.org/10.1007/978-3-030-28715-3_3

in the sense that it was difficult to teach, and in the sense that I had to do something to overcome the challenge. I had to develop a teaching method, which would allow using the "standard" body of knowledge on standards, while making the teaching interesting for students and gratifying for myself. Besides, differences in learning outcomes had to be considered, depending on the course and the department, where the course has been taught.

In the following, I share my experience and knowledge on teaching about standards to students of different majors. In doing so, I attempt to show the interdependence of topics and concepts, the "tips and tricks" of didactic to turn the dry concept of standards and standardization from boring to intriguing in the eyes of students. I start from sketching my own body of knowledge on standards in order to establish a common ground with the reader.

2 View on Standards Through the Eyes of the Lecturer

Standards have long been acknowledged to play an important role in local and global competition (Funk, 1998, 2002), national and international industrialization policy-making (Fomin, Pedersen, & de Vries, 2008; Kwak, Lee, & Fomin, 2011), strategy and new product development (Farrell & Saloner, 1985), among others. At the turn of the millennia, standards were held responsible for 80% of the global trade, or $4 trillion annually (OECD, 1999, p. 4).

There are many definitions available for the word "standard" and different types of standards can be distinguished. A technical standard is an established norm or requirement in regard to technical systems. It is usually defined in a document as a set of specifications for manufacturers to adhere to (David & Greenstein, 1990, p. 4). The word "standard" can also refer to methods, processes and practices, in which case the word "norm" can be used instead.

Among the most familiar to general public standards are *A4* (in Europe) and *Letter* (in the U.S.) office paper sizes—the standard[1] prescribes specific characteristics (the dimensions) of the paper sheet, but does not prescribe the manufacturing method. A standard can also do the contrary—prescribe a method for doing something without imposing any specific characteristics of the product being subjected to that method, as e.g., EPA standards[2] for testing for presence of harmful substances in the air.

While less known or visible, compatibility standards (Farrell & Saloner, 1985; West & Fomin, 2011) are of ultimate importance in the contemporary networked society—defining both product characteristics and working methods to enable inter-connection and inter-operability of electronic and informational tools. These standards can be as simple as those establishing inter-connectivity (defining substitutes) for electricity or telecommunications wall sockets and plugs, and as complicated as those establishing inter-operability of different mobile cellular handsets and services

---

[1] https://en.wikipedia.org/wiki/ISO_216.

[2] https://www.epa.gov/measurements/collection-methods.

across cellular networks around the world. In educational context, the International Organization for Standardization's (ISO) **OSI interoperability model**[3] is often used to demonstrate the breadth of issues addressed by compatibility standards in the world of Information and Communication Technologies (ICT).

Standards typically originate from standard setting organizations (SSOs)[4] or industry consortia (Farrell & Saloner, 1988; Funk & Methe, 2001; Schmidt & Werle, 1998), where interested stakeholders develop a standard through specified consensus mechanisms. Standards, which are developed and released by formal bodies like ISO are referred to as de jure standards (Folmer & Verhoosel, 2011, p. 18), meaning that the standards come from a recognized legal entity (hence—made according to the law, or "de jure") and that the development of the standard took place according to the specified rules for reaching consensus. De facto standards can be released by industry consortia or any kind of organization, including private companies. In the latter case, only when the proprietary standard is widely used it starts to be referred to as de facto standard.

"De jure" is often (and in my view quite misleadingly) associated with mandatory nature of a standard—i.e., when the use of a standard is required by law. However, de jure standards are not mandatory by definition. For example, national authorities can set de facto or de jure standard for the local market players to adhere to (comply with) on a compulsory basis. An example of **mandatory** de jure standard can be electricity current's voltage in a de-monopolized utilities market. An example of **mandatory** de facto standard can be the use of Adobe's PDF (Portable Document Format) in national e-government initiatives (Fomin et al., 2008) prior to 2005, when PDF became a de jure standard as ISO 19005-1:2005, or the use of Windows or Linux operating system in government offices. While imposing mandatory standards in national or international markets is a common practice (Pedersen, Fomin, & de Vries, 2009), sometime market forces prove to be stronger than the governing will of the authority, leading to non-adherence or failure to impose a mandatory standard (see e.g., Kwak et al., 2011). An example here can be the recent "dieselgate" scandal,[5] when Volkswagen automaker had intentionally programmed diesel engines to circumvent the strict restrictions of Clean Air Act.

In the contemporary deregulated global economies, de jure standards often compete against other de jure standards, as well as against de facto standards (Besen & Farrell, 1994; Farrell & Saloner, 1987; Funk & Methe, 2001). The latter refers to products, methods or services, which achieved high market recognition nationally or internationally, but which have not been specified by any SSO or standards consortia. It is not uncommon for de facto standards to be taken to a relevant SSO in order to "promote" their status to de jure. A well known example of de facto made into de jure standards are Microsoft Office document formats (DOC, XLS, PPT), which were originally developed as proprietary company products, have seen

---

[3]https://en.wikipedia.org/wiki/OSI_model.

[4]https://en.wikipedia.org/wiki/Standards_organization.

[5]https://en.wikipedia.org/wiki/Volkswagen_emissions_scandal.

global acceptance—i.e., became de facto standards—and only then were passed on to formal SSOs for obtaining the status of "official" de jure standard.

Promoting proprietary technologies to the status of de jure standards may open new national and international markets to the promoting company (Pedersen et al., 2009; Suttmeier & Yao, 2004). However, the company which is successful in establishing it's own technologies as de facto standard may be content with keeping it this way. For example, there are many de facto standards found in the product portfolio of Apple Inc.—the company has a long tradition of developing proprietary technologies and products and benefitting from exclusive rights on them.

Establishing a desired standard, or securing specific favorable properties to be included in the standard under development, more often than not is a daunting challenge for any single player, particularly because: (1) there are many market players with diverse and competing interests (Farrell & Saloner, 1988) with a potential to get involved in the standardization process; (2) local firms often lack knowledge and expertise on technology standardization (Barlette & Fomin, 2008; de Vries, Fomin, & Dul, 2009); and (3) because of the high cost of participation in international standardization (Jakobs, Procter, & Williams, 1998). Due to the aforementioned challenges, large multinational corporations are better positioned to succeed in establishing their proprietary technologies as de facto or de jure standards. This, however, does not preclude small enterprises or even individuals from succeeding in this endeavor, as witnessed by such standards as e.g., MP3 music codec and the Linux operating system. Needless to say that along with the global recognition, the developers are likely to receive substantial direct or indirect rents, usually due to the intellectual property rights (IPRs) or patents associated with the standard (Blind & Jungmittag, 2008; Blind & Thumm, 2004; Iversen, 1999; Updegrove, 2007b).

Rents and economies of scale associated with standards establish the significance of standardization in the development of national industries, necessitating national governments to promote and protect the interests of local firms in the international standardization arena (Fomin et al., 2008; Funk, 2000; Su & Fomin, 2010; Suttmeier & Yao, 2004). Governments may help domestic firms acquire innovation capability by promoting state-led consortia for technology development (Lyytinen & Fomin, 2002; West & Fomin, 2001), establishing science parks, intervening in royalty negotiations with firms in developed countries or driving standardization for national technologies. While some companies and governments alike have long recognized an axiom that the lack of standards, or the existence of unfavorable standards is likely to create obstacles for market development and exports (Kwak et al., 2011; Schmidt & Werle, 1998; Updegrove, 2007a), the majority of firms may still lack that awareness.

Given the significance of standards and standardization on firm, national, and global levels, there has been a substantial growth in the body of knowledge in this field. Some disciplines may have benefited more than others from this growth. Educational programs in technical universities, educational programs on international trade—those are more likely beneficiaries of the growing popularity of the topic. Within the fields I teach in—Informatics and Organizational Management—standards and standardization research has remained marginal, with percentage volume

From Boring to Intriguing: Personal Perspective on Making ...

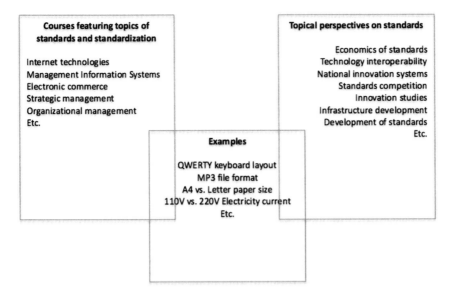

Fig. 1 The challenge for the lecturer in teaching about standards is to find a good match between the subject area, the theoretic perspective, and the examples which can intrigue students

measured in single digits (Lyytinen & King, 2002). At the same time, the last decade saw a growing effort by well recognized SSOs and national standards bodies (NSBs) in promoting education about standardization through mainstream educational programs.

How to approach the task of educating about standardization, if the teaching subjects are not those where the bulk of standardization research is found? Or where the standards are understood as narrowly as technical specification or a communication protocol, as it is often the case for the students of Informatics? The task is complicated not only by the "dry" or "boring" tags typically attached to standardization topics by students, but also by the fact that standardization (standard making process) can be approached from different theoretical perspectives. In this situation, the challenge for the lecturer is to find a good match between the subject area, the theoretic perspective (if relevant), and the examples to use (see Fig. 1).

3 Teaching on Standards and Standardization: How to Turn "Boring" into "Intriguing"

3.1 Informatics and Information Systems Studies

Incorporation of standards-related topics into technical disciplines may be expected as a "natural" thing to do. What comes as unexpected is the degree of ignorance of students on anything related to standards. One of the courses I have taught to undergraduate students at the Faculty of Informatics is "Internet Technologies"[6] under the major of Multimedia and Internet Technologies, in which my task as a co-lecturer was to introduce the topic of standards and standardization.

Students majoring Multimedia and Internet Technologies must obtain a decent understanding of the fact that technology standards form the core of Internet and are *sine qua non* for interoperability of multimedia applications and devices. Hence, knowledge on technology standards must be a compulsory part of the program's curricula. What I have learned from teaching the students of Informatics—the two things which have not changed over the 4 years of this course taught—first, that the 3rd year undergraduate students are mostly ignorant about standards, to the extent that they cannot comprehensively define what a standard is, where does it come from, not even mentioning students' ability to tell a de jure from de facto standard (and the implications of this difference in the context of the Internet and Multimedia applications), and, second, how easy it is to intrigue students for learning more if proper examples are given as a teaser.

In my first year's teaching, I tried to intrigue students by giving them "the big picture"—the role of standards in the Internet and the global economy. I quickly came to realize that this method did not work. Students are rarely interested in "big" things, in the "macro" order of the world. Phrases like "standards are responsible for 80 percent of the world trade" make students yawn, at best. They better perceive the "micro" picture of the world—their immediate world, as constituted by their everyday activities and "things" they use on a daily basis.[7] As students of Informatics, they also have passion for "developing things"—software, user interfaces, and even devices. That aspect became the foundation of my new teaching strategy—I focused on "things" which students use or develop.

"Have you ever downloaded music in MP3 format?", "Have you ever shared your MP3 files with friends?", "Have you been able to 'play' MP3 on different devices?", "Do you know who developed MP3?", "Do you know why MP3 can be called a standard?" … If the first three questions ultimately yield "yes" from the students, the last two have always produced "no"—the anchoring has been made, now students realized they did not know something about the thing they thought is so familiar to them, and they are ready to learn (about) this boring term "standard" just to restore their confidence in "educated self" and fix the "familiarity bond" to the thing they thought they knew so well.

Through the "anchoring to the familiar thing", a "black box" of a standard can be opened. The well known "MP3" alias for music files is demystified—the **history** of the creation of the standard, the **competition** in becoming established as de facto world digital music format, the work of people on **establishing** it as **de jure standard** and **maintaining** the standard, the role of the **producers** of the equipment and the **consumers** making choices in purchasing the equipment and downloading the music,

---

[6]http://if.vdu.lt/en/studies/bachelor-studies-programmes/multimedia-and-internet-technologies/.

[7]For a scholarly discussion on the different scales in how the world, the Internet, etc. can be perceived (and studied) I recommend (Edwards, 2003).

the role of Internet **applications developers** in creating **support** for the format and ensuring **interoperability** across devices, sites, and services—all those topics can be explored at once or one by one at ease, as students are now eager to learn how a boring technical word "standard" can mean **so many different new things** for the "dull thing" they thought they knew so well—MP3.

The new knowledge on standards comes as a treasure box to students trained to think of algorithms, software, and the design process. Having grasped the notions pertaining to standards and standardization—de facto and de jure, path dependence and standards wars, standards ownership and installed base—students can better understand the concept of "Internet technologies"—that's the name of the course I teach—as a living socio-technical monster, which—students suddenly realize—is fed not only by technical people developing software code and companies manufacturing hardware devices, but also **by students themselves through their choices of which standards to support and which not** in designing and using multimedia applications for the Internet. The otherwise foreign and dull topic of "standards wars" (dull, because for students of informatics anything coming from economists is always #boring #dull #irrelevant) now can be seen in a different light—each software design or use case by students contributes to the ongoing war craft.

In teaching the module of the standards within the "Internet technologies" course, I introduce to the students of Informatics technology-, organization-, and economics-related topics. The organizational prism on the development of standards gives the students knowledge of the institutional infrastructure of standards development. With this knowledge, students can tell the difference between de jure and de facto standards, which in turn helps them make better informed choices when working with one or another multimedia format, designing user interfaces, etc. For example, when asked about the standout of Adobe Flash versus HTML5 in Apple products, students usually know the technical aspect of the case—Apple does not readily embrace support for Adobe Flash for the reasons of higher computation power (and hence battery power) requirements. Having learned about SSOs and proprietary company formats, students can look at those two and other multimedia formats they know through a different lens. Formerly purely technical considerations in choosing which technology to support, now can be given more versatile analysis, which should hopefully lead to better design choices in the future.

The "technology" aspects of standards I teach about, are those related to the standards themselves—what the standard is, what does it specify or relate to. Having learned about SSOs and the national standardization body (NSB), students are asked to explore the online catalogs of those organizations, find multimedia-related standards, preferably those they use on a daily basis or have heard about, and get familiar with the content of the document containing a description of the standard. A typical unexpected (for students) learning outcome of this exercise is when students realize how many there are different standards related to their field of interest. Having grasped the breadth of the issues covered by domain-specific standards, students receive normative knowledge on formal standards development with appreciation. Now they realize also that **their innovative ideas** can be brought to National Standardization Body (NSB) and eventually materialize into a useful standard one day.

The economic aspects of standards are taught using the concepts of market competition (also referred to as "standards wars") and **network externalities** (also referred to as **network effects**) (Katz, 1986; Katz & Shapiro, 1985). Here, again, I use the same teaching strategy—give students a "thing" they believe they know very well and then intrigue them with a teaser in the form of an unknown interesting fact about that "familiar thing". Why do I use the concept of network externalities as a proxy for "economics of standards"? For students of Multimedia and Internet Technologies program, understanding the market mechanisms behind the interoperability of services is of paramount importance. They do learn about application program interfaces (APIs), how to develop and use them to establish interoperability. But that technical knowledge cannot help them understand why Internet application developers and users choose one or another application or service. Why Apple does not readily support Adobe Flash format? Why YouTube has several times changed the default encoding standard for videos it hosts? By having a more versatile understanding of what drives company and consumer choices, the future app developers are better prepared to make wiser choices on their side—which services/applications to support in the apps they develop.

Internet is a rich source of stories demonstrating the concept of network externalities. "Who does not have a Facebook account?", "Why do you use Facebook?", "When did you open Facebook account and why?", "Would you use Facebook if you were the only user?"… By answering those simple questions students, indeed, help the lecturer explain the concept of network externalities. This generation of students still remembers "the times before Facebook" and the social media sites they used prior to switching to Facebook[8]—drawing on students own experience (in this case—experience in the process of and motivation for switching from one product to another) is **a method** I find giving good learning outcomes.

To explain how strong (or persisting) the network effects can be, I use another example—that of QWERTY keyboard layout.[9] Again, this is the "thing" every student is familiar with. Rarely, if at all, know the students the story behind the layout and why the attempts to introduce less awkward (and more efficient) layouts—e.g., DVORAK—have been a failure so far. The "thing" which was so much familiar to students can now be seen as something completely different from what they had always believed it was—from a handy computer input device it becomes a "lame device" and the standards—as seen through the prism of network externalities—are to blame for this.

The QWERTY example is also helpful in demonstrating imperfection of technical standardization, incompleteness of technical view on standards in understanding the use- and user-related sides of technology. It is also helping explain the **path**

---

[8]In Lithuania, prior to Facebook being established as a dominant social media application, there were several popular social media sites there: www.one.lt, www.klase.lt, etc. Today those sites still exist, but seem to be used by minor user groups for specific (unknown to the author) purposes. When asked whether still using those sites, students give negative answer.

[9]There are many urban legends found on the Internet about the QWERTY layout. I use Stephen Jay Gould's "The Panda's Thumb: More Reflections in Natural History" as a motivational source.

dependency concept, which, in turn, can help explain how and why **standards must be maintained**.

3.2 Organizational and Management Studies

If I was to teach on economics of standards, or on institutional setup of national and global levels of standardization, that would be a "straight forward" thing to do in a Master students' (graduate) class of International Business or Business Management programs. My goals are slightly different, and the audience is slightly different, too. Since 2013, I teach a course "Business management and decision making",[10] which is elective to students of different programs of the Faculty of Economics and Management. Up to 75% of students in this course are visiting foreign students of different majors, which makes the diversity of students' interests and their background knowledge even greater.

My goal is to teach not on technical standards, but on **organizational norms and processes**—i.e., **standard operational procedures** (SOPs) or **routines** as foundation of organizational management and decision making processes. In this respect, the knowledge students receive could be applied in the context of analysis of e.g., **management system standards** (MSS), such as ISO/IEC 9001-series Quality Management System standard, ISO/IEC 27001-series Information Security management standards, etc. MSSs are directly related to company's working practices (Barlette & Fomin, 2009; Bousquet, Fomin, & Drillon, 2009) and require certain knowledge not only about the subject matter (quality, information security, etc.), but also about the organizational processes and the role of employees in supporting or not supporting (e.g., sabotaging) those processes.

My teaching goals are twofold here. Not only do I have to introduce the concept or standard operational procedures, but, more importantly, explain the students that when dealing with organizational norms, biases and workarounds are commonplace. I.e., from day one of the course I am trying to establish that the classical Wiberian approach to organizational management is good insofar as it explains how organization should work (this is similar to the concept of a reference standard prescribing a method), but it is more important for managers to understand why people, processes, and organizations **often do not work as expected** (de Vaujany, Fomin, Lyytinen, & Haefliger, 2018), and what can be done to minimize non-conformance to established norms or management standards and the risk of organizational failure. In other words, my goal is to explain that often there is a great divide between "should" and "is" in organizational management, explain that norms and management standards do not

---

[10]http://www.vdu.lt/en/studies/courses/faculty-of-economics-and-management/, http://www.vdu.lt/wp-content/uploads/2017/05/Management-and-Business-Decisions-VAV5018.pdf.

only come from the management (from the top), but may also be emerging from the daily working practices.[11]

My teaching strategy focused on illuminating and elaborating **non-conformance** to organizational norms, the concept of the "norm" as seen through the proxy of the concept of "routine" (Feldman & Pentland, 2003) helps prepare managers capable of appreciating the possibilities offered by management standards in boosting efficiency and effectiveness of organizational conduct, in minimizing different organizational risks stemming from the human conduct. By the separation of technical-, organizational-, and individual/behavioral aspects of norms and standard operating procedures I attempt to establish the versatility of the concept of norm, and the need for managers to deal equally well with the different perspectives on norms, understand **how norms are developed, maintained, changed**. As the reader can see, the issues addressed are exactly the same as those to be addressed in teaching about technology standards.

3.3 Emerging Topics for Education About Standardization

Conducting teaching at two different departments—to students of Informatics and to students of Economics and Management, I also see some common denominators in the body of standards and standardization knowledge. I believe this knowledge will be helpful in delivering teaching about standardization in the next five to 10 years. Those emerging topics for education about standardization are related to the global economic and societal developments in general, and the role of technologies in business and private life in particular. The concept of **Industry 4.0**, or **the 4th industrial revolution**, shows the importance of having a broad, versatile knowledge on standards and standardization—the one capable of "connecting the dots" of technical interfaces and organizational norms, for example. The example from the study of McKinsey company (2015), which demonstrates the case of loss of 99% of data in informing the company's operational management is the one I often use in the class (see Fig. 2).

On the one hand, we need technology (interoperability) standards to collect and communicate data to drive the 4th industrial revolution. On the other hand, we need appropriate organizational management practices (norms, routines, standard operating procedures) to take advantage of the capabilities offered by contemporary technology. And the other way around—if managers expect to receive the promised benefits of Industry 4.0, a good understanding of technology standards will come handy in setting organizational change strategies. Having only one of the two perspectives on standardization—the technical or the organizational—is likely to result in mismanagement, as demonstrated by McKinsey report.

---

[11]The reader may notice, that the concepts of top-down and bottom-up development of organizational norms are very close to the concepts of de jure and de facto standards.

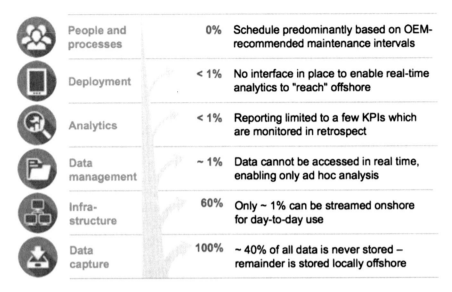

Fig. 2 Case example of 99% of all data lost before reaching operational decision makers. Measurements made at oil rig with a base of 30,000 data tags. *Source* McKinsey (2015)

Other emerging topics where education about standardization will be important, are those of e.g., **cyber-human systems** (European Commission, 2013; Schoitsch & Niehaus, 2017)—the introduction of technologies into lives of people as part of rehabilitation, assistive living, prosthetics, etc.—and **robot law** (Reuters Staff, 2017)—the introduction of Artificial Intelligence in judicial system. The number of university programs to embrace the topic of standards will increase accordingly to include Health and Social Care, Law, Ethics and Philosophy, among other. What is more important, those emerging socio-technical paradigms call for the revision of standards for standardization—i.e., for establishing new standards on how standards are made. **Responsible standardization (RS)** is a term which has been acquiring a growing importance recently, emphasizing that standards have to deliver not only on technical expectations, but also meet societal needs and follow ethical norms of users of the technologies (who are often absolutely ignorant about the standards those products incorporate).

4 Conclusions

This chapter was intended to share the author's experience in education about standardization. The major challenge I came across in teaching about standards and standardization is that it is difficult to instigate students' interest in the topic if teaching about standards per se—what a standard is and how or where it is created. Given the challenge, I had to develop an "intriguing strategy" first. At the heart of this teaching strategy (or method) is the use of examples related to things or concepts familiar to students, whereby the examples are used to show the students that contrary to what they believed, there are knowledge gaps in their "familiarity", and that filling those gaps requires learning about standards and standardization. Put it simple, my approach to education about standardization is to demonstrate the usefulness of the concept of standards prior to explaining the concept itself.

With a personal experience of teaching about standards at two different departments, I came to realize how much students appreciate good examples on standards and standardization, and how difficult it may be for a lecturer to find good examples. Despite the growing interest to education about standardization, to date there is scarcity of inter-disciplinary, non-normative, and interactive educational materials on standardization—the kind of educational resources, which can intrigue students, and which education about standardization is in dire need today.

The blurring boundaries between technological, economic, legal, ethical, etc. perspectives on standards and standardization, as outlined in this chapter, calls or new methods and strategies in research and education about standardization, and for sharing experiences and resources among the educators. Maintaining a narrow perspective on standards in teaching will not cater well for educating competent and responsible specialists.

It is my sincere hope that through my teaching, my students can acquire the foundation for the necessary knowledge base and analytical breadth to take on the roles of protagonists and game changers in solving contemporary and future organizational and societal challenges related to or stemming from the dull word "standard", and that my colleagues in academia can find my personal approach to education about standardization useful in enriching and improving their own.

References

Barlette, Y., & Fomin, V. V. (2008). *Exploring the suitability of IS security management standards for SMEs* (pp. 1–10). Hawaii: IEEE. https://doi.org/10.1109/HICSS.2008.167.

Barlette, Y., & Fomin, V. V. (2009). The adoption of information security management standards: A literature review. In *Cyber-security and global information assurance: Threat analysis and response solutions* (pp. 119–140). Hershey, PA: IGI Global. Retrieved from http://www.igi-global.com/chapter/adoption-information-security-management-standards/7413.

Besen, S. M., & Farrell, J. (1994). Choosing how to compete: Strategies and tactics in standardisation. *Journal of Economic Perspectives, 8,* 117–131.

Blind, K., & Jungmittag, A. (2008). The impact of patents and standards on macroeconomic growth: A panel approach covering four countries and 12 sectors. *Journal of Productivity Analysis, 29,* 51–60.

Blind, K., & Thumm, N. (2004). Interrelation between patenting and standardisation strategies: Empirical evidence and policy implications. *Research Policy, 33,* 1583–1598.

Bousquet, F., Fomin, V. V., & Drillon, D. (2009). *Standardization: A major tool for competitive intelligence* (pp. 27–36). Cergy, France: Wissenschaftsverlag Mainz in Aachen.

David, P. A., & Greenstein, S. (1990). The economics of compatibility standards: An introduction to recent research. *Economics of Innovation and New Technology, 1,* 3–41.

de Vaujany, F.-X., Fomin, V. V., Lyytinen, K., & Haefliger, S. (2018). Research commentary—Rules, practices and information technology (IT): A trifecta of organizational regulation. *Information Systems Research* (in print).

de Vries, H. J., Fomin, V., & Dul, J. (2009). *International standardization in a multi-stakeholder environment—How a stakeholder can pursue influence.* Barcelona, Spain.

Edwards, P. N. (2003). Infrastructure and modernity: Force, time, and social organization in the history of sociotechnical systems. In *Modernity and technology* (pp. 185–226). Cambridge, MA: MIT Press.

Egyedi, T. M. (2007). *IEC lecture series: The importance of standards.* Geneva: IEC, CD-ROM.

European Commission. (2013). *Cyber-physical systems: Uplifting Europe's innovation capacity* (https://ec.europa.eu/digital-agenda/en/cyber-physical-systems-1). Unit A3-DG CONNECT, Communications Networks, Content & Technology Directorate-General.

Farrell, Joseph, & Saloner, G. (1985). Standardization, compatibility, and innovation. *Rand Journal of Economics, 16,* 70–83.

Farrell, J, & Saloner, G. (1987). Competition, compatibility and standards. In *Product standardization and competitive strategy.* Amsterdam: Elsevier Science Publishers.

Farrell, Joseph, & Saloner, G. (1988). Coordination through committees and markets. *The Rand Journal of Economics, 19,* 235–252.

Feldman, M. S., & Pentland, B. T. (2003). Reconceptualizing organizational routines as a source of flexibility and change. *Administrative Science Quarterly, 48*(1), 94–118. https://doi.org/10.2307/3556620.

Folmer, E., & Verhoosel, J. (2011). *State of the art on semantic IS standardization, interoperability and quality.* Retrieved from www.semanticstandards.org.

Fomin, V. V., & Keil, T. (2000). *Standardization: Bridging the gap between economic and social theory* (Vol. Paper 20, pp. 206–217). Brisbane, Australia: AIS. Retrieved from http://aisel.aisnet.org/icis2000/20.

Fomin, V. V., Pedersen, M. K., & de Vries, H. J. (2008). Open standards and government policy: Results of a Delphi survey. *Communications of the Association for Information Systems, 22,* 459–484.

Funk, J. L. (1998). Competition between regional standards and the success and failure of firms in the world-wide mobile communication market. *Telecommunications Policy, 22,* 419–441.

Funk, J. L. (2000, September 26–27). *The mobile internet market: Lessons from Japan's i-mode system.* Presented at the E-Business Transformation: Sector Developments and Policy Implications, Washington, DC.

Funk, J. L. (2002). *Global competition between and within standards. The case of mobile phones.* Palgrave.

Funk, J. L., & Methe, D. T. (2001). Market- and committee-based mechanisms in the creation and diffusion of global industry standards: The case of mobile communications. *Research Policy, 30,* 589–610.

Iversen, E. (1999). *Standardization and intellectual property rights: ETSI's controversial search for new IPR-procedures.* Presented at the IEEE Conference on Standardisation and Innovation in Information Technology, Aachen, Germany. Retrieved from https://eprints.utas.edu.au/1297/1/Iversen_ETSI_2OO2.pdf.

Jakobs, K., Procter, R., & Williams, R. (1998). User participation in standards setting—The panacea? *ACM StandardView, 6,* 85–89.

Katz, M. L. (1986). Technology adoption in the presence of network externalities. *Journal of Political Economy, 94,* 822–841.

Katz, M. L., & Shapiro, C. (1985). Network externalities, competition, and compatibility. *American Economic Review, 75,* 424–440.

Kwak, J., Lee, H., & Fomin, V. V. (2011). The governmental coordination of conflicting interests in standardisation: Case studies of indigenous ICT standards in China and South Korea. *Technol-*

ogy Analysis & Strategic Management, 23(7), 789–806. https://doi.org/10.1080/09537325.2011.592285.

Lyytinen, K., & Fomin, V. V. (2002). Achieving high momentum in the evolution of wireless infrastructures: The battle over the 1G solutions. *Telecommunications Policy, 26,* 149–170. https://doi.org/10.1016/S0308-5961(02)00006-X.

Lyytinen, K., & King, J. L. (2002). Around the cradle of the wireless revolution: The emergence and evolution of cellular telephony. *Telecommunications Policy, 26,* 97–100.

McKinsey. (2015). *Industry 4.0: How to navigate digitization of the manufacturing sector.* McKinsey & Company. Retrieved from https://www.mckinsey.de/files/mck_industry_40_report.pdf.

OECD. (1999). *Regulatory reform and international standardization* (Working Party of the Trade Committee No. TD/TC/WP(98)36/FINAL). Paris: Organization for Economic Cooperation and Development. Retrieved from https://www.oecd.org/tad/benefitlib/1955309.pdf.

Pedersen, M. K., Fomin, V. V., & de Vries, H. J. (2009). The open standards and government policy. In *ICT standardization for E-business sectors: Integrating supply and demand factors* (pp. 188–199). Hershey, PA: IGI Global. Retrieved from http://www.igi-global.com/chapter/information-communication-technology-standardization-business/22931.

Reuters Staff. (2017, February 16). *European parliament calls for robot law, rejects robot tax.* Retrieved from https://www.reuters.com/article/us-europe-robots-lawmaking/european-parliament-calls-for-robot-law-rejects-robot-tax-idUSKBN15V2KM.

Schmidt, S. K., & Werle, R. (1998). *Coordinating technology. Studies in the international standardization of telecommunications.* Cambridge, Massachusetts: The MIT Press.

Schoitsch, E., & Niehaus, J. (2017). *Strategic agenda on standardization for cyber-physical systems.* CP-SETIS. Retrieved from ISBN 978-90-817213-3-2.

Su, J., & Fomin, V. V. (2010). Balancing public and private interests in ICT standardization: The case of AVS codec. In *What kind of information society? Governance, virtuality, surveillance, sustainability, resilience*, IFIP Series (Vol. 328/2010, pp. 64–72). Springer. Retrieved from http://www.springerlink.com/content/q4wrv42046743876/.

Suttmeier, R., & Yao, X. (2004). *China's post-WTO technology policy: Standards, software, and changing nature of techno-nationalism* (No. Special Report No. 7). The National Bureau of Asian Research. Retrieved from http://www.nbr.org/publications/issue.aspx?ID=126.

Updegrove, A. (2007a). Government policy and "standards—based neo-colonialism." *Standards Today. A Journal of News, Ideas and Analysis, VI,* 1–9.

Updegrove, A. (2007b). It's time for IPR equal opportunity in international standard setting. *Standards Today. A Journal of News, Ideas and Analysis, VI,* 1–3.

West, J., & Fomin, V. V. (2001). *When government inherently matters: National innovation systems in the mobile telephony industry, 1946–2000.* Presented at the Academy of Management annual meeting, Washington, DC. Retrieved from http://aom.org/Meetings/annualmeeting/Past-Annual-Meetings.aspx.

West, J., & Fomin, V. V. (2011). Competing views of standards competition: Response to Egyedi & Koppenhol. *International Journal of IT Standards & Standardization Research (JITSR), 9,* i–iv. https://doi.org/10.4018/IJITSR.

Prof. Vladislav V. Fomin holds a position at Vilnius University in Kaunas, Lithuania and a principal researcher position at Turiba University in Riga, Latvia. Prof. Fomin also holds visiting professor's positions at the Faculty of Computing of the University of Latvia and La Rochelle Business School in France. From 2009 to 2019, Vladislav was a professor at Vytautas Magnus University in Kaunas, Lithuania. Vladislav Fomin has over 100 scientific publications in journals, conferences, and as book chapters, including Information Systems Research, Academy of Management Journal, Journal of Strategic Information Systems, Telecommunications Policy, and other. A substantial part of Vladislav's research is found within the domain of standards and standardization.

Mutual Enforcement of Research and Education—The Case of Structured Inquiry-Based Teaching of Standardization

Geerten van de Kaa

1 Introduction

Envision a situation in which a teacher starts his or her lecture by posing various research questions to the students. Students then tackle these questions by discussing them and gradually an answer emerges. However, the answer leads to new questions that are subsequently discussed. This form of inquiry-based teaching has been used for thousands of years and is very effective in training students' analytical skills. As many students seem to like this type of teaching (Van de Kaa, 2013), the author has used inquiry-based in his elective course *Technology Battles* offered at the Technology Policy and Management Faculty of Delft University of Technology. This chapter explores to what extent research, education, and practical relevance are related to each other and focuses on the context of education and research about standards battles. Thus, the case of education on standards battles is an illustrative case that serves the goal of assessing whether the relation holds.

This paper discusses the advantages of structured inquiry in teaching. It thus builds on the research on structured inquiry (Banchi & Bell, 2008; Colburn, 2000) in general and its benefits for standardization education specifically (Van de Kaa, 2013). It contributes to the research on standardization education by providing further empirical evidence of the relationship between education and research on standardization that was first introduced in Van de Kaa (2013). It also contributes to that research by including practical relevance as a separate concept that is linked to both research and education.

It has been chosen to apply structured inquiry in the teaching, but it must be emphasized that there are other types of inquiry-based teaching such as guided and open/true inquiry and types of teaching that do not adhere to the principles of inquiry-

---

G. van de Kaa (✉)
Delft University of Technology, Delft, The Netherlands
e-mail: g.vandekaa@tudelft.nl

© Springer Nature Switzerland AG 2020
S. O. Idowu et al. (eds.), *Sustainable Development*, CSR, Sustainability,
Ethics & Governance, https://doi.org/10.1007/978-3-030-28715-3_4

based teaching. However, we believe that structured inquiry is most applicable to teaching at a MSc. level as we believe it leads to more independent learning.

Section 2 presents a model linking research, education and practical relevance. In Sect. 3, we apply the model to a case and present various illustrations of the workings of the model. We discuss extant research on standards battles, since this is the subject of the case described in this paper. Section 4 comprises a discussion and conclusion consisting of theoretical recommendations, practical implications, limitations, and areas for future research.

2 Framework

Faculty at universities basically have two tasks. First, they are involved in teaching. They teach courses at various levels and supervise M.Sc. students and or Ph.D. candidates in writing their theses and dissertations respectively. Second, they are involved in research. Our first proposition is that these two tasks should be intertwined. Our second proposition is that both education on standardization and research on standardization should have practical relevance. In the remainder of this section we will describe these two propositions in more detail.

2.1 Research and Its Influence on Education

Research can have an influence on education in various ways. First, research in the form of scientific papers or book chapters that are published by key authors in the field may be used as required reading at university. Second, own publications may be used as required reading for lectures that are offered by faculty members. Students are encouraged to critically review these publications before class and the teacher can explain them in depth during the lectures.

2.2 Education and Its Influence on Research

Education can also lead to research in several ways. First, sometimes students come up with novel ideas and they may choose to apply those ideas in their own projects such as in an M.Sc. thesis research project. Although this might not occur very often during the early stages of the student's academic teaching trajectory (B.Sc. level), at later stage in their studies, excellent students may transform their theses into publications. Second, students may participate in large scale surveys and the data gathered may be used as empirical data. These empirical data may be used by faculty in their own research to test certain hypotheses. This is often done during courses offered to large groups of B.Sc. students. Third, assignments that are written by students or data that is collected by students may be used as empirical data to develop new theory or test existing hypotheses. However, it is doubtful whether data collected by students is of a sufficient quality.

2.3 Education and Research Have Practical Relevance

It is evident that education on standardization has practical relevance. Through teaching, students' knowledge and understanding of a certain topic is increased and their analytical skills are improved. Students can subsequently apply this acquired knowledge and these skills in industry. Students often do an internship as part of their B.Sc. or M.Sc. thesis trajectory in which case teaching (writing a thesis) has practical relevance. This thesis may be read by practitioners, increasing their knowledge and understanding of the particular issue.

Most research provides insights that are relevant for practice as well. Many papers offer a separate section that discusses policy or management relevance. Some journals such as the California Management Review and the Harvard Business Review publish practical versions of scientific papers. These journals are read by practitioners in industry and may include practical insights that are relevant to them.

2.4 A Framework Linking Research, Education, and Practical Relevance

These arguments result in a framework that links research, education, and practical relevance (see Fig. 1). Academic research leads to student work through inquiry-based teaching. Student work can subsequently be used in research projects. Both academic research and teaching have practical relevance.

3 Teaching Academic Research for Practical Relevance

We describe the elective course *Technology Battles* which serves as an illustration of the relationships depicted in Fig. 1. This course is offered in the Management of Technology programme that is offered at the Faculty of Technology, Policy, and Management at Delft University of Technology, the Netherlands. The programme is open to students from outside the Faculty and from abroad. About half of the students are of Dutch origin and come from the other technological faculties and higher educational institutions. The other half comes from other countries such as Indonesia, China, and Finland. Thus, the population is quite diverse.

Fig. 1 Framework linking academic research, teaching, student work, and practical relevance [partly based on Van de Kaa (2013)]

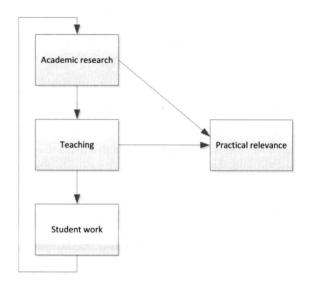

This course is about standards battles. First, we provide a general overview of the research on this topic to show the theoretical background of the course. Many scholars have studied high tech markets in which technology is developed which increases in value the more users adopt that technology (Gallagher, 2012; Gallagher & Park, 2002; Schilling, 2002). These markets are characterized by increasing returns to adoption. Network effects (Farrell & Saloner, 1985; Katz & Shapiro, 1985) arise in such markets, and often the winner takes all (Hill, 1997). This means that such markets often result in a single dominant design or standard. Practitioners and scholars alike are increasingly interested in understanding the mechanisms that underlie these characteristics (Sheremata, 2004). Which factors explain standard success? Various scholars have developed frameworks to determine the outcome of standards battles (Den Uijl & De Vries, 2008; Schilling, 1998; Suarez, 2004; Van de Kaa, Van den Ende, De Vries, & Van Heck, 2011) and these frameworks have been applied to various cases to understand the relevance and applicability of these factors (Gallagher, 2012; Gallagher & Park, 2002; Garud & Kumaraswamy, 1993; Van de Kaa & De Vries, 2015). Practitioners may apply the frameworks as checklists to try to decrease the uncertainty when choosing for a standard (Van de Kaa et al., 2011). These decisions are crucial as betting on the wrong standard can result in substantial financial resources being spent on developing and marketing a standard that will not be used in the end. For example, some years ago a battle was fought between HD-DVD and Blu-ray. Toshiba developed and promoted the HD-DVD standard which turned out to be unsuccessful and the company lost a huge amount of money as a result (Gallagher, 2012).

Van de Kaa et al. (2011) developed a list of 29 factors for standards success, divided into five categories. The categories include (1) complementary assets in the form of e.g. financial resources, (2) strategies that may be pursued such as penetration pricing, (3) technological characteristics such as ensuring a certain level of backwards compatibility (Schiavonea, 2014), (4) market mechanisms such as network effects, the level of uncertainty in the market, and, path dependencies (Arthur, 1989) and, (5) other stakeholders such as the regulator.

Various scholars have attempted to understand which factors are crucial for various cases. For example, factors for standard dominance have been applied to various cases including energy systems (Shapiro & Varian, 1999; Van de Kaa, Kamp, & Rezaei, 2017; Van de Kaa, Rezaei, Kamp, & De Winter, 2014; Van de Kaa, Scholten, Rezaei, & Milchram, 2017), ICT systems (Den Hartigh, Ortt, Van de Kaa, & Stolwijk, 2016; Heinrich, 2014; Van de Kaa, De Vries, & Rezaei, 2014; Van de Kaa, Van Heck, De Vries, Van den Ende, & Rezaei, 2014; Vanhaverbeke & Noorderhaven, 2001), consumer products (Schilling, 1999, 2003; Srinivasan & Venkatraman, 2010) and telecommunications (Funk, 1998, 2002). Additionally, scholars have focused on specific factors for standard success including installed base and the availability of complementary goods (Schilling, 1999, 2003), marketing communications (Dranove & Gandal, 2003), commitment (Kamps, De Vries, & Van de Kaa, 2017; Van de Kaa, Van den Ende, & De Vries, 2015), effectiveness of the standard development process (Van de Kaa & De Bruijn, 2015), and structure of the industry wide network of standards organizations (Vanhaverbeke & Noorderhaven, 2001). Finally, some scholars have attempted to explain why multiple standards co-exist for some standards battles (De Vries, De Ruijter, & Argam, 2011).

3.1 General Description of the Course

In the course, students analyze a standards battle and determine the importance of factors for standard success for that particular case. After the course, students should be able to understand the theoretical background of standards battles and they should be able to write a scientific research paper on the subject.

The course is organized into six regular sessions and follows the inquiry-based teaching method whereby research and teaching is perfectly aligned. Each session consists of a combination of presentations and discussions. Students are required to prepare for these sessions by studying and analyzing various assigned scientific articles. They must answer various questions including:

- What type of paper is this (descriptive, conceptual, literature review/meta-analysis, theory development, theory testing, etc.)?
- What academic perspective does the paper take?
- What are the theoretical arguments/propositions?
- What is the conceptual model, if any (independent variable, dependent variable, moderating variable etc.)?
- What is the research methodology?
- What are the results and how does this paper contribute to the field of study?
- In what way is it valuable (or not) for you or your fellow students?

- Based upon this paper, what are the next steps (recommendations for further research?).

In other words, students are asked to perform all the usual academic steps that are taken in a research project that is conducted in the social sciences, and they are also explicitly asked to be critical towards the theory.

These questions are later discussed in class. The teacher and the students will further deepen their understanding of the topics and themes presented in each paper. All students are expected to participate in the class-wide discussion. This is done in a classical way whereby the teacher raises research questions and the students tackle these questions by engaging in academic discussion and applying the knowledge gained from reading the assigned articles.

3.2 Assignment

All students in the *Technology Battles* course must complete an assignment that is developed according to the principles of structured inquiry (Banchi & Bell, 2008; Van de Kaa, 2013). Students are required to write a scientific paper in which they examine a standards battle by analyzing the factors that affected the outcome of the standards battle and by determining the weights of these factors by applying the Best Worst Method (BWM). Excellent papers may be submitted to a publication outlet. Students that wish to specialize in standards write a master's thesis on the topic by applying the empirical data gathered in the course.

Students work in small groups to complete the assignment. Although the teacher will provide examples of standards battles, the group must choose a specific standards battle. Ideally, the group will choose a case in which the group members have some expertise.

Students begin by analysing the existing literature on the standards battle and determine which factors were mentioned in the publication. Maybe the literature on the specific battle does not mention all factors for standard success. This could mean that these factors do not apply to the case or that adding one or more of these factors could provide a better explanation of the case. Next, students collect secondary data in the form of reports, practitioners' literature, external databases, etc. which they use to write a preliminary version of the case description. At this stage, they need to conduct additional interviews with practitioners to fully reconstruct the case. They should conduct at least three interviews per standards battle. A pre-defined questionnaire, posted by the teacher on an online learning environment, can be used when conducting interviews. Interviews should be taped and transcribed. Quotes should be used to back up claims made in the report. At this stage, students can determine which standard achieved dominance and which factors were relevant for standard dominance. A factor is relevant when it is mentioned explicitly or implicitly in the secondary sources (literature) or primary sources (interviews).

During their analysis of the standards battle, students should be able to link all of their observations back to the literature (so when a particular factor appears to be particularly relevant, this should be explained by making use of primary or secondary sources).

The assignment is limited to 7000 words. Papers should be short, compact, and clear. Papers are graded on:

1. Quality of the argumentation
2. Structure of document and arguments
3. Theory-based reasoning
4. Fact (data) based reasoning
5. Compactness
6. Clarity of argumentation
7. General understanding of the topic

The data part of the paper is not limited in size and is graded on:

1. Clarity of presentation
2. Completeness (within reasonable boundaries)
3. Validity [e.g., did the student consider multiple sources (including interviews)]
4. Possibility of triangulation.

3.3 The 2016–2017 Course

The 2016–2017 course was offered to 24 students, who were divided in eight groups. Some students continued to write a master thesis on the topic that they chose during the course and some are now even writing a paper based on their thesis. Some papers can be used as required reading material in future courses. However, most graduates quickly get a job in industry and often do not complete their paper.

3.4 Evaluation of the Course in Light of the Framework

Our case study provides several illustrations of the relationships depicted in Fig. 1. First, research in the form of publications are used in education. Furthermore, student assignments sometimes become publishable papers. Also, students sometimes write M.Sc. theses that stem from the teaching that was provided. Thus, this paper provides further empirical evidence of the possible relationship between education and research [thus building on Van de Kaa (2013)].

Students that have followed the course on *Technology Battles* have gained considerable knowledge about factors for standard success and can apply such factors in their future jobs. Some students will find employment at major consultancy firms and can advise firms that are uncertain about which standard to promote. Other students might have the opportunity to work at a multinational such as Philips that have a standardization department and might apply the factors directly to various cases in which the company is involved. In both cases the uncertainty attached to choosing for a standard will decrease. Students might also find work at a regulatory institution or a standardization agency and apply the knowledge gained.

4 Discussion and Conclusion

This paper describes the two-way relationship between research and education and offers examples of how education on standardization may lead to research and to practical relevance.

4.1 *Practical Feedback from Students*

Student evaluations of the course were overall quite positive. Students indicated that they enjoyed this type of teaching but that it was challenging in terms of analyzing the literature and collecting data. Students should be aware that they are required to identify literature and data sources at the start of the course and that they should allocate a significant part of their time to collecting data. They can only draw meaningful conclusions if they have read all (or a significant part of) the literature that reports on the case study. Although the number of hours budgeted for this course is equal to the number of hours that would have to be invested according to the EC system, students still find this course very difficult.

4.2 *Contributions and Implications*

We contribute to the literature on education on standardization (Van de Kaa, 2013) by providing new empirical proof for the relationship between education and research. Furthermore, we extend the model presented in Van de Kaa (2013) by including practical relevance as a concept and we explain the various ways in which research and education have practical relevance.

We recommend that academics should structure their courses according to the principles of structured inquiry and that they adhere to the relationship between education and research as much as possible. We believe this may increase the quality of education as teachers are constantly updating required reading based on the most recent publications available. Required learning material that includes contributions from former students may encourage students to work harder on assignments. The rationale behind this is that students do not complete assignments just to receive a pass grade, but that the end result of their efforts might result in a publishable paper. This might offer them more motivation to work on the assignment. This is a proposition that could be empirically studied in future research.

4.3 Limitations and Recommendation for Future Research

The limitation of this study is that we have studied only one case of academic teaching on standards battles. The successful application at Delft University of Technology does not imply that the approach we suggest is the best for all other situations Future research should study more cases of academic teaching on other topics inside and outside the field of standardization which might give more support to our claims.

In this chapter, level 2 inquiry-based teaching (structured inquiry) is applied throughout the course described. This means that students are given a detailed description of the research tasks that should be pursued. This is described in detail in Sect. 4.2. The question is whether similar results would be obtained at higher levels of inquiry-based teaching. For example, in level 3 inquiry-based teaching (guided inquiry) a teacher only provides a research question. In this case, students are responsible for the research needed to answer the question but can follow their own approach. Then, other methodologies than interviewing may be used to determine the importance of factors for standard success. Alternatively, level 4 inquiry-based teaching may be applied (open/true inquiry) where students formulate research questions on their own. Future research could examine whether similar results can be obtained when following these forms of education. For the course described in this chapter, both level 3 and level 4 inquiry-based teaching may be more difficult and more applicable to courses that are offered at a graduate school level where Ph.D. candidates are required to conduct research projects on their own. This requires a certain amount of creativity on the part of students, but it is certainly viable.

References

Arthur, W. B. (1989). Competing technologies, increasing returns, and lock-in by historical events. *The Economic Journal, 99*(394), 116–131. Retrieved from http://links.jstor.org/sici?sici=0013-0133%28198903%2999%3A394%3C116%3ACTIRAL%3E2.0.CO%3B2-R.
Banchi, H., & Bell, R. (2008). The many levels of inquiry. *Science and Children, 46*(2), 26–29.
Colburn, A. (2000, March). An inquiry primer. *Science Scope,* 42–44.
De Vries, H. J., De Ruijter, J. P. M., & Argam, N. (2011). Dominant design or multiple designs: The flash memory card case. *Technology Analysis & Strategic Management, 23*(3), 249–262.
Den Hartigh, E., Ortt, J. R., Van de Kaa, G., & Stolwijk, C. C. M. (2016). Platform control during battles for market dominance: The case of Apple versus IBM in the early personal computer industry. *Technovation, 48–49,* 4–12.
Den Uijl, S., & De Vries, H. (2008). *Setting a technological standard: Which factors can organizations influence to achieve dominance?* Paper presented at the 13th EURAS Workshop on Standardisation.

Dranove, D., & Gandal, N. (2003). The DVD versus DIVX standard war: Empirical evidence of network effects and preannouncement effects. *Journal of Economics and Management Strategy, 12*(3), 363–386.

Farrell, J., & Saloner, G. (1985). Standardization, compatibility, and innovation. *The Rand Journal of Economics, 16*(1), 70–83.

Funk, J. L. (1998). Competition between regional standards and the success and failure of firms in the world-wide mobile communication market. *Telecommunication Policy, 22*(4/5), 419–441.

Funk, J. L. (2002). *Global competition between and within standards—The case of mobile phones.* Basingstoke UK/ New York, NY, USA: Palgrave.

Gallagher, S. R. (2012). The battle of the blue laser DVDs: The significance of corporate strategy in standards battles. *Technovation, 32*(2), 90–98.

Gallagher, S. R., & Park, S. H. (2002). Innovation and competition in standard-based industries: A historical analysis of the U.S. home video game market. *IEEE Transactions on Engineering Management, 49*(1), 67–82.

Garud, R., & Kumaraswamy, A. (1993). Changing competitive dynamics in network industries: An exploration of sun microsystems' open systems strategy. *Strategic Management Journal, 14*(5), 351–369. Retrieved from http://links.jstor.org/sici?sici=0143-2095%28199307%2914%3A5%3C351%3ACCDINI%3E2.0.CO%3B2-I.

Heinrich, T. (2014). Standard wars, ties standards, and network externality induced path dependence in the ICT sector. *Technological Forecasting & Social Change, 81*, 309–320.

Hill, C. W. L. (1997). Establishing a standard: Competitive strategy and technological standards in winner-take-all industries. *Academy of Management Executive, 11*(2), 7–25.

Kamps, X., De Vries, H., & Van de Kaa, G. (2017). Exploring standards consortium survival in high tech industries: The effects of commitment and internal competition. *Computer Standards & Interfaces, 52,* 105–113.

Katz, M. L., & Shapiro, C. (1985). Network externalities, competition, and compatibility. *American Economic Review, 75*(3), 424–440. Retrieved from http://links.jstor.org/sici?sici=0002-8282%28198506%2975%3A3%3C424%3ANECAC%3E2.0.CO%3B2-M.

Schiavonea, F. (2014). Backwards compatibility, adapter strategy and the 'battle of converters' in analogue photography. *Technology Analysis & Strategic Management, 26*(4), 401–416.

Schilling, M. A. (1998). Technological lockout: An integrative model of the economic and strategic factors driving technology success and failure. *Academy of Management Review, 23*(2), 267–284. Retrieved from http://links.jstor.org/sici?sici=0363-7425%28199804%2923%3A2%3C267%3ATLAIMO%3E2.0.CO%3B2-F.

Schilling, M. A. (1999). Winning the standards race: Building installed base and the availability of complementary goods. *European Management Journal, 17*(3), 265–274.

Schilling, M. A. (2002). Technology success and failure in winner-take-all markets: The impact of learning orientation, timing, and network externalities. *Academy of Management Journal, 45*(2), 387–398.

Schilling, M. A. (2003). Technological leapfrogging: Lessons from the U.S. video game console industry. *California Management Review, 45*(3), 6–32.

Shapiro, C., & Varian, H. R. (1999). The art of standards wars. *California Management Review, 41*(2), 8–32.

Sheremata, W. A. (2004). Competing through innovation in network markets: Strategies for challengers. *Academy of Management Review, 29*(3), 359–377.

Srinivasan, A., & Venkatraman, N. (2010). Indirect network effects and platform dominance in the video game industry: A network perspective. *IEEE Transactions on Engineering Management, 57*(4), 661–673.

Suarez, F. F. (2004). Battles for technological dominance: An integrative framework. *Research Policy, 33*(2), 271–286. Retrieved from http://www.sciencedirect.com/science/article/B6V77-49SFH5C-1/2/6ac467f816758fde3d35b8edf195c27b.

Van de Kaa, G. (2013). Structured inquiry and standardization. *International Journal of Innovation, Management and Technology, 4*(2), 233–237.

Van de Kaa, G., & De Bruijn, J. A. (2015). Platforms and incentives for consensus building on complex ICT systems: The development of WiFi. *Telecommunication Policy, 39,* 580–589.

Van de Kaa, G., & De Vries, H. (2015). Factors for winning format battles: A comparative case study. *Technological Forecasting and Social Change, 91*(2), 222–235.

Van de Kaa, G., De Vries, H. J., & Rezaei, J. (2014a). Platform selection for complex systems: Building automation systems. *Journal of Systems Science and Systems Engineering, 23*(4), 415–438.

Van de Kaa, G., Kamp, L. M., & Rezaei, J. (2017a). Selection of biomass thermochemical conversion technology in the Netherlands: A best worst method approach. *Journal of Cleaner Production, 166,* 32–39.

Van de Kaa, G., Rezaei, J., Kamp, L., & De Winter, A. (2014b). Photovoltaic technology selection: A fuzzy MCDM approach. *Renewable and Sustainable Energy Reviews, 32,* 662–670.

Van de Kaa, G., Scholten, D., Rezaei, J., & Milchram, C. (2017b). The battle between battery and fuel cell powered electric vehicles: A BWM approach. *Energies, 10,* 1707–1720.

Van de Kaa, G., Van den Ende, J., & De Vries, H. J. (2015). Strategies in network industries: The importance of inter-organisational networks, complementary goods, and commitment. *Technology Analysis & Strategic Management, 27*(1), 73–86.

Van de Kaa, G., Van den Ende, J., De Vries, H. J., & Van Heck, E. (2011). Factors for winning interface format battles: A review and synthesis of the literature. *Technological Forecasting and Social Change, 78*(8), 1397–1411.

Van de Kaa, G., Van Heck, H. W. G. M., De Vries, H. J., Van den Ende, J. C. M., & Rezaei, J. (2014c). Supporting decision-making in technology standards battles based on a fuzzy analytic hierarchy process. *IEEE Transactions on Engineering Management, 61*(2), 336–348.

Vanhaverbeke, W., & Noorderhaven, N. G. (2001). Competition between alliance blocks: The case of RISC microprocessor technology. *Organization Studies, 22*(1), 1–30.

Geerten van de Kaa is Associate Professor of Standardization and Business Strategy at Delft University of Technology. He holds a Ph.D. from Rotterdam School of Management, Erasmus University. His research interests include platform wars for complex systems; (collaboration) strategies for innovation; energy systems; and (responsible) innovation and standardization. He is the author and co-author of more than 100 publications. He has published in international journals including *Organization Studies, IEEE Transactions on Engineering Management, Technovation, Technological Forecasting and Social Change, Telecommunications Policy, Technology Analysis & Strategic Management, Renewable and Sustainable Energy Reviews, Journal of Cleaner Production, R&D Management, Computer Standards & Interfaces,* and *Journal of Systems Science and Systems Engineering.* He teaches courses in the strategic management of technological innovation.

Addressing Sustainability in Education About Standardisation—Lessons from the Rotterdam School of Management, Erasmus University

Henk J. de Vries

1 Introduction

Solving societal issues such as poverty, inequality, the aging population and climate change requires cooperation between many actors. Potential solutions are often systemic, meaning that they involve interrelated processes, products and services. Producing these solutions creates new business opportunities but the issues also question current business practices from which the societal issues may have resulted. This creates a tension. The system-character of possible solutions implies that no single stakeholder is in the position to develop and implement the system, that close cooperation between private and public stakeholders is needed, and that common standards should be developed to enable the creation, functioning and further evolution of the systems. This makes standardisation an important instrument for addressing sustainability issues.

The most commonly accepted set of goals for sustainable development are the United Nations Sustainable Development Goals (SDGs). If indeed standardisation is essential for achieving these goals, then standardisation should be included in education programmes for sustainability, as it is in the master *Standardization, Social Regulation and Sustainable Development* at the University of Geneva (Université de Genève, 2018). This paper starts, however, from the standardisation education side by describing, in Sect. 2, the design of the standardisation course curriculum at the *Rotterdam School of Management, Erasmus University* (see for an earlier description de Vries, 2005). This attempt to design a well-balanced multidisciplinary master elective on standardisation and innovation has lead to a curriculum in which sustainability is included. Section 3 shows how the SDGs are addressed in this course.

H. J. de Vries (✉)
Rotterdam School of Management, Erasmus University, Rotterdam, The Netherlands
e-mail: hvries@rsm.nl

Faculty of Technology, Policy and Management, Delft University of Technology, Delft,
The Netherlands

© Springer Nature Switzerland AG 2020
S. O. Idowu et al. (eds.), *Sustainable Development*, CSR, Sustainability,
Ethics & Governance, https://doi.org/10.1007/978-3-030-28715-3_5

Chapter 4 discusses the lessons that can be learnt from this case on integrated teaching of standardisation and sustainability.

2 Education About Standardisation

2.1 *Introduction to the Course Theme*

Rotterdam School of Management, Erasmus University (RSM) is the business school of the (public) Erasmus University. University rankings list RSM as one of Europe's top 10 business schools. RSM has 6000 bachelor and master students, 400 MBA students, and per year 2200 executives. The department *Technology and Operations Management*'s section on *Innovation Management* has an endowed chair (professorship) on standardisation, partly funded by the Netherlands Standardisation Institute NEN. RSM offers 15 full-time one year master programmes such as *Strategic Management*, *Supply Chain Management*, and *Management of Innovation*. Each master programme includes core courses, elective courses and a master thesis trajectory. One of the elective courses in the *Management of Innovation* master is *Innovation and Interface Management*. Students are expected to spend 168 h on this course, during seven weeks. The term interface replaces the earlier term standardisation in order to make the course more appealing to students. This name change indeed made more students choose this elective. And the change can be justified as by definition, standardisation has to do with interfaces: standardisation is the activity of establishing and recording a limited set of solutions to actual or potential matching problems, directed at benefits for the party or parties involved, balancing their needs, and intending and expecting that these solutions will be repeatedly or continuously used, during a certain period, by a substantial number of the parties for whom they were meant (de Vries, 1997). These matching problems concern interfaces between entities, or for an entity because of its interfaces with other entities. So standards provide stable solutions related to interfaces. Most standards describe such a solution or provide performance requirements or test methods (de Vries, 1998).

As a business school, RSM aims to educate future managers. Standards are not necessarily on the agenda of board meetings of companies. So what should future managers know? From a company's perspective, the main strategic standardisation questions are:

- What should we do standard, what tailor-made? In general, 'standard' is cheaper, but customer wishes differ, so not everything should be standardised, how to find the right balance?
- In the case of 'standard': do we make our own specification or do we use an external standard?
- In the latter case, do we take our external environment for granted, or do we want to influence it, e.g., by participation in external standardisation?

The other core concept is innovation. The combination of standardisation and innovation is not self-evident, the two concepts even seem to exclude each other. Standardisation has to do with 'freezing' a solution for a matching problem for a certain period of time, whereas innovation has to do with something new—in the classic definition of Schumpeter (1934), innovation is the commercialisation of all new combinations based upon the application of new materials and components, the introduction of new processes, the opening of new markets and/or the introduction of new organisational forms. If standards freeze a solution, that solution cannot be changed anymore until the standard is withdrawn or revised so in that sense it indeed hinders innovation. However, for interoperability purposes standard interfaces are needed and then the interface remains stable whereas the interconnected modules can be further innovated (Lee, 2010). In the case of performance standards or measurement standards the 'technical' solutions are not determined. Literature gives evidence that standardisation supports rather than hinders innovation, see for an overview of studies the recent *Handbook of Innovation and Standards* (Hawkins, Blind, & Page, 2017). This makes standardisation a necessary topic in an innovation management master.

Most participants in the course already had core courses in innovation management. Therefore, the emphasis in the elective is on interface management/standardisation management as such, and on its integration in innovation management. Designing a curriculum relates to contents but also to an educational approach. We begin with the latter.

2.2 *Educational Approach*

de Vries (2015) argues that for proper understanding of standardisation a combination of different scientific angles is needed, offered both by fundamental and by applied sciences and that business research is the discipline needed to integrate these contributions from other disciplines because standardisation is primarily a business activity. Similar arguments apply for innovation. So a business school is a proper place for a standardisation currilum and the approach should be multidisiplinary. van de Lagemaat (1986) developed such a multidisciplinary approach. I chose to build on his work also because his approach is founded in the same philosophy (Dooyeweerd, 1955, 1957) I used in a study mapping the research field of standardisation (de Vries, 2015) and in a study on (management systems) standards application (Haverkamp & de Vries, 2016).

According to van de Lagemaat (1986, p. 28), business education has a twofold purpose: disclosing the subject matter for students, and disclosing students for the subject matter. Then the purpose of standardisation education is to disclose the standardisation phenomenon in a way the students can understand it and to act with students in a way that they get accustomed with standardisation, get knowledge about it and are equipped for using this knowledge in business practice. In academic teaching the focus should be on critical thinking rather than on 'how to' recipes and tools.

In van de Lagemaat's educational approach (1986, p. 183), four questions are leading: *what* is done in practice (in this case, by people involved in standardisation), *how* is it done, *why* is it done, and *why* is it done *in this way*? Standardisation management does not prepare for all activities but rather for steering these for the benefit of the company and other stakeholders, and in relation to innovation management.

A problem in academic education is that due to their research activities, many researchers no longer see reality as it is, but 'see' mainly the aspects related to their own discipline. This, of course, influences their education activities. In order to avoid such one-sidedness, van de Lagemaat (1986) recommends to pay systematic attention to the diversity of aspects of phenomena and to use exemplary learning (real-life case studies) to demonstrate how the different aspects apply. This leads to a better understanding of reality. Real-life examples appeal not only to the intellectual capacities of a student but to the student as a person as well, so they are of help in opening the students for standardisation. They will discover that standardisation cannot be understood by a 'rationalistic' technical and economical approach only, but that standards are developed and applied by interacting people with all sorts of habits. Of course, examples are not enough—there should be theory too, so that, through the example(s), the phenomenon as such becomes clearer.

In the course this is done at the very first day in the form of a role playing game, developed by Tineke Egyedi of (at that time) Delft University of Technology (now: DIRoS, Amsterdam) and the company United Knowledge. This case is about an innovative technology for which an interface is essential, and standardisation of some aspects of the interface is needed. Each student gets the role of a certain stakeholder and they have to negotiate in a standardisation committee. Though it is a game and not reality, it makes them familiar with core issues that will be addressed in more detail later during the course. Company visits and guest lectures are another format to make students familiar with business practice and a visit to the national standards body NEN complements this. Company hosts and guest lecturers (e.g. former Ph.D. candidates of the chair who now work in industry) are chosen based on their ability to interconnect their business practice to academic theory. Due to practical constraints, the number of visits had to be reduced to two in 2019, and the number of guest lectures to two. In lectures and scientific papers the diversity of scientific angles is provided. Then in assignments students have to integrate this and make sense of what they have learnt. In a personal assignment they have to give a solution for a problem they have experienced themselves with a human interface, and in a group assignment they have to advise to one of the companies they have visited. Finally they have to write an essay on a self-chosen topic related to innovation and interface management in which they apply scientific theories and provide a critical discussion.

2.3 Dimensions of Standardisation to Be Covered in the Course

The course aims to make students familiar with standardisation and my intention is not to expose the phenomenon in its broadness, thus including all relevant dimensions. Verman (1973) distinguished three dimensions:

1. Subject (engineering, transport, housing/building, food, agriculture, forestry, textiles, chemicals, commerce, science, and education);
2. Aspect (nomenclature and symbols, specification, sampling and inspection, tests and analysis, grading and classification, simplification and rationalization, code of practice and byelaws, packaging and labelling, and forms and contracts);
3. Level (individual, company, association, national, and international).

However, even more dimensions can be distinguished:

1. the company, intercompany, local, national, regional and global level;
2. different business sectors;
3. different subject matter areas, technical as well as non-technical ones;
4. different categories of standards;
5. different processes of standardisation (see Fig. 1);
6. different stakeholders and their stakes;
7. different aspects of standards and standardisation.

Including everything in a curriculum seems impossible: each dimension has several options so any combination of six options constitutes a possible standardisation topic, and each topic can be studied using different research angles, the 7th dimension. However, in practice it may be doable, as:

1. the company, inter-company, local, national, regional and global level are not totally different—broadly speaking the processes are similar;
2. differences in categories of standards remain and include more diversity than Verman's listing suggests (de Vries, 1998; Egyedi & Ortt, 2017) but the processes related to developing and applying them do not differ fundamentally, so in fact these differences do not have to be treated as an extra 'dimension' but can be a separate topic in the standardisation course;
3. standardisation differs per business sector, but these are practical differences, not fundamental ones and in some sectors, not all kinds of standardisation apply;
4. the same applies to subject matter areas: practical differences, no fundamental ones;
5. the processes all should be addressed;
6. the stakeholders can be distinguished in stakeholders within a company (e.g., marketing, purchasing, quality management) and those in society (e.g. producers, professional users, consumers, workers) and their associations, and these need to get attention but the role of these in standardisation shows similarities as well (e.g., trade associations and consumer organisations face similar barriers in getting involved in standardisation).

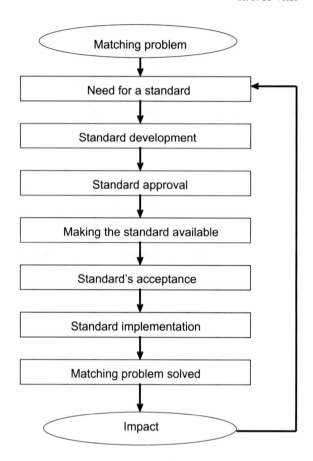

Fig. 1 Simplified process model of standardisation (de Vries, 2008)

For the design of the curriculum this means that different combinations of dimensions have to be made, and linked to innovation. To mention an example: one of the papers students have to read addresses the impact of energy performance standards on innovation in the construction sector (de Vries & Verhagen, 2016). This paper addresses (1) the intercompany and national levels, (2) a combination of a compulsory performance standard and a voluntary measurement standard, (3) a business sector (construction) that can be characterised as a loosely coupled system (Dubois & Gadde, 2002) and in that sense is comparable to and thus informative about other sectors such as shipbuilding, entertainment, education and healthcare, (4) the topic of energy use, (5) the processes of standards development (by a national government and in committees of a national standardisation institute), standard use (by architects and construction companies) and impacts (on technical and organisational innovation, on CO_2 reduction and on the country's competitive position), and (6) stakeholders including consumers, architects, construction companies, municipal authorities, national government and a research organisation. Technical innovations include ventilation and heating technologies, and these innovations can be incremental or systemic. The latter also lead to innovations at the level of groups of houses, process innovations (related to IT support, e.g. design software), and organisational innovation (new forms of cooperation within the sector).

The next step is to relate the different research angles to the topics, to make the course multidisciplinary. The main research angles are offered by the fundamental sciences, each of these sciences studies one of the aspects distinguished by Dooyeweerd (1955, 1957): the arithmetic, spatial, kinematical, physical, biotic, feelings-, logical, historical, linguistic, social, economic, aesthetic, juridical, ethical and faith aspect. Related to each aspect there are 'laws' that should be honoured. For instance, mathematical laws of adding, subtracting, multiplying, etc., apply to the arithmetic aspect and are given. For the logical and later aspects, these 'laws' are normative principles that man, in his freedom, can obey, ignore or oppose, e.g., economic laws.

Haverkamp and de Vries (2016) applied this aspect approach for analysing feelings of uneasiness related to in-company process standardisation. The aspects approach served both as a checklist in order not to forget aspects that might be relevant, and as a normative tool of how to do things 'better,' in that case to improve process efficiency via standardisation without causing uneasiness.

It can be disputed if these aspects should be presented in a systematic way to students as advised by van de Lagemaat (1986). It is a philosophical approach that is not specific to standardisation and may be applied everywhere. Another option is to leave this approach to the teachers when developing a curriculum, this is what I do. The different scientific disciplines can be related to the fifteen aspects (de Vries, 2015) and their findings should be used in a multidisciplinary standardisation curriculum. The aspects themselves do not constitute the curriculum but most aspects deserve to get specific attention, in particular the economic aspect because standardisation is an economic activity (Blind, 2004; de Vries, 1997), the law aspect because standards are the result of an agreement (Fickers, Hommels, & Schueler, 2008), the linguistic aspect because the standard is a tool for communication expressed in language (plus figures and tables) (Teichmann, 2010), (mostly) physical aspects related to the matching problem for which the standards provides a solution, and the social aspect because people cooperate in creating the standard (Lelong & Mallard, 2000). But also, for instance, the arithmetic aspect gets some attention by presenting the topic of preference ranges (Bongers, 1980), including a small in-class exercise about preference ranges in kite surfing, and the faith aspect by discussing a paper on different forms of conformity assessment—do these enhance or even affect trust?

2.4 Presenting Topics in the Form of Managerial Dilemmas

RSM being a management school, the course contents should be managerial. Therefore topics are presented as managerial dilemmas. The main ones are (1) balancing between 'standard' and tailor-made, and (2) the related theme of standardisation versus innovation. Additionally the, also related, dilemmas of (3) cooperation versus competition, (4) public versus private, (5) global versus local, and (6) standard versus authentic are included in the curriculum. Next, it is explained *how* standardisation can be used to achieve company goals: (1) variety management, (2) developing and implementing company standards, (3) meeting external standards, (4) conformity assessment, (5) selecting the standard that is expected to win, and influencing the dominance battle, (6) successful participation in (international) standardisation, (7) impacts of standards and standardisation, (8) designing an architecture of standards, (9) governance of standardisation in a company, and, last but not least, (10) how to solve societal issues using standardisation.

All is based on research at RSM (therefore, many self-written sources are used), and is related to the current academic debate (therefore, most academic literature is recent) and to issues in current standardisation practice (e.g. the new law in standardisation in China, the Belt and Road Initiative, recent international trade agreements, and the problems with Europe's success story, the New Approach) and business news (e.g. platforms connecting two- or multisided markets, the shared economy, circular economy, autonomous driving, 3D printing, the Chinese import of milk-powder, the diesel engine scandal, recent mergers and acquisitions).

2.5 Resulting Course Programme

The above provides the standardisation-side of the course on standardisation and innovation management. As mentioned, most students are familiar already with innovation management but the different forms of innovation should be addressed in their relation to standardisation, and the integration of standardisation in the innovation process is a core topic (Wiegmann, 2016). The latter also includes the timing of standardisation: anticipatory, participatory or responsive (Sherif, 2001).

After the introduction game the course starts with some simple examples of standards related to the life world of students, such as McDonald's and smart phones. At a later moment, business-to-business examples are added. After these examples, generalisation to the concepts of standards and standardisation is possible, including general advantages of standardisation, definitions, decision making in standardisation, different types of standards, and different ways to arrive at standards.

The next step is to explain (formal) national standardisation. This is no longer the most important level of standardisation. Yet we start with it, as it is the easiest one to explain: easier than standardisation at regional, global, company or intra-company level. After having discussed the national level, it is relatively easy to discuss formal standardisation at the global level: ISO and IEC. ITU differs slightly from ISO and IEC and this is just mentioned without going into any details. The European level can be added after the national and global one, as knowledge of both is necessary to understand it. In a similar way, CEN and CENELEC get more attention than ETSI. Then the Chinese and the US systems are explained and the strengths and weaknesses of the different systems.

Once the formal standardisation has been explained, it is possible to pay attention to consortium standardisation. After this it is easy to mention other forms, such as sectoral standardisation. Next, standardisation within companies can be discussed. As this includes the company use of external standards and the company's strategy in influencing external standardisation, it should not be on an earlier place in the curriculum. Knowledge of company standardisation is necessary before the subsequent topic can be discussed: intra-company standardisation. Etc. Once the basics are clear it is possible to focus on the managerial dilemmas and the *how to*. Table 1 shows an overview of the course.

The lectures are in the first four weeks of this seven weeks course, except for 11 and 12. During these weeks students also have to make the human interface assignment (10% of the course grade). Week 5 starts with a written exam (30%) and at the end of that week some of the students present their human interface case to fellow students. The last weeks are for finishing the group assignment (advice to the company Interface Europe Manufacturing, 20%) and the essay (40%). The high percentage for the essay shows the importance of critical thinking, the main assessment criterion for this assignment. Reflecting the pedagogical approach, the course starts broad by letting students experience reality as it is, then it delves into

Table 1 Master elective *Innovation and Interface Management* 2018—overview of lectures

Lecture	Articles
1. Introduction; game: setting standards for innovation	–
2. Interface management—introduction and history	de Vries (2008)
3. Stakeholder analysis; system approach to standardisation	de Vries (2018)
4. **Guest lecture 1**: platforms	Eisenmann, Parker, and Van Alstyne (2006)
5. In-company standardisation; services innovation	de Vries (2006) and de Vries and Wiegmann (2017)
6. **Guest lecture 2**: the emergence of de-facto standards	Van de Kaa, van den Ende, de Vries, and van Heck (2011), den Uijl and de Vries (2013), de Vries, de Ruijter, and Argam (2011)
7. Management system standardisation; conformity assessment	Manders, de Vries, and Blind (2016), de Vries, Feilzer, Gundlach, and Simons (2010), Haverkamp and de Vries (2016)
8. Modes of standardisation; impact of standardisation on innovation; patents	Wiegmann, de Vries, and Blind (2017), Bekkers (2017), den Uijl, Bekkers, and de Vries (2013)
9. **Company visit**: responsible innovation	–
10. Standardisation and international trade; authenticity and standardisation; responsible innovation	de Vries, Nagtegaal, and Veenstra (2017), de Vries and Go (2017), de Vries and Verhagen (2016)
11. Human interface	Dul et al. (2012)
12. Discussion group assignment course wrap-up	

the details of a multitude of theoretical lenses, and finally it broadens again, both in a practical sense (give an integral advice to a real-life company), and more theoretical (the essay).

3   Addressing Sustainability

3.1   The UN Sustainable Development Goals

Sustainability is often conceptualised in terms of the 'triple bottom line' (Elkington, 1999), meaning that companies should help society achieve economic prosperity, environmental protection as well as social equity. Other definitions are used as well, sustainability may 'refer to social responsibility, ethics, or a larger piece of the strategic management rubric and has also been tied to strategic decision-making. It also has ties to ecological concerns, the natural environment and (…) economic and social concerns' (Bateh, Horner, Broadbent, & Fish, 2014, p. 35). Among practitioners the most common definition of sustainable development is 'development that meets the needs of the present without compromising the ability of future generations to meet their own needs' (World Commission on Environment and Development, 1987). At the international political level, sustainability goals have been approved in 2015. A description of these SDGs can be found in the report *Transforming our World: the 2030 Agenda for Sustainable Development* (United Nations General Assembly, 2015). In 2015, countries adopted the 2030 Agenda for Sustainable Development and its 17 Sustainable Development Goals.

'Over the next fifteen years, with these new Goals, countries will mobilize efforts to end all forms of poverty, fight inequalities and tackle climate change, while ensuring that no one is left behind. The SDGs, also known as Global Goals, build on the success of the Millennium Development Goals (MDGs) and aim to go further to end all forms of poverty. The new Goals are unique in that they call for action by all countries, poor, rich and middle-income to promote prosperity while protecting the planet. They recognize that ending poverty must go hand-in-hand with strategies that build economic growth and addresses a range of social needs including education, health, social protection, and job opportunities, while tackling climate change and environmental protection. While the SDGs are not legally binding, governments are expected to take ownership and establish national frameworks for the achievement of the 17 Goals. Countries have the primary responsibility for follow-up and review of the progress made in implementing the Goals, which will require quality, accessible and timely data collection. Regional follow-up and review will be based on national-level analyses and contribute to follow-up and review at the global level' (United Nations, 2018).

3.2 Addressing the UN Sustainable Development Goals in RSM's Standardisation Course

For RSM, the SDGs serve as a reference framework with an implicit link to what the School does when it comes to research and education. The standardisation course was designed before this decision was taken. Therefore, sustainability is not included in the learning goals of the course, but as this section will show: most of the subsequent goals do get attention in the master elective *Innovation and Interface Management*.

Goal 1. End poverty in all its forms everywhere.
Standardisation is a core instrument in eliminating non-tariff barriers to trade. In general, but not always, this contributes to more prosperity. A study on the Transatlantic Trade and Investment Partnership TTIP (de Vries et al., 2017) reveals how such non-tariff barriers to trade can be removed not only between the US and Europe but also elsewhere in the world. However, standards in the North may also make it difficult for companies in developing countries to export. This is being told using a case described in a master thesis on the impact of European food safety standards on the export of shrimps from Bangladesh to the EU (Langemeijer, 2004). Some lessons from a recent study on the relation between standardisation and innovation in developing countries are being mentioned as well (Lee, Zoo, & de Vries, 2017).

Goal 2. End hunger, achieve food security and improved nutrition and promote sustainable agriculture.
The above applies to Goal 2 as well. Food safety management is taken as a case to trace and classify stakeholders related to standardisation. Additionally, certification seals related to sustainability and fair trade are mentioned, and own research shows how western consumers react to such seals (de Vries, 2016).

Goal 3. Ensure healthy lives and promote well-being for all at all ages.
Again, certification seals are of help. Ergonomics, including ergonomic standards, is relevant in this sense as well (Karwowski, 2006). Students get an assignment on how to improve a human interface, using a paper on ergonomics (Dul et al., 2012), and an in-class assignment on the role of standards for the ageing population in Japan.

Goal 4. Ensure inclusive and equitable quality education and promote lifelong learning opportunities for all.
The historical example of guilds is mentioned—standards and conformity assessment were used to get well-educated craftsmen at local level. Also the impacts of guilds on innovation are discussed (Epstein & Trak, 2008).

Goal 5. Achieve gender equality and empower all women and girls.
Apart from mentioning that standardisation committees tend to be dominated by men I do not address gender issues. In discussing ISO 9001 and the business excellence models from Japan, the USA and Europe I point at the difference of seeing employees as real human beings (as in the European EFQM model) or as 'just' human resources (ISO 9001 and Baldridge criteria; the Japanese model is in-between).

Goal 6. Ensure availability and sustainable management of water and sanitation for all.
I illustrate the importance of standards for testing by referring to a master thesis project in Indonesia (Steenvoorden, 2008). Drinking water from rivers has to be purified first. This process depends on the amount of mud in the water, which fluctuates due to (sometimes extreme) rainfall. The Technical University of Bandung developed a method to measure the amount of mud. Knowing this amount allows to make purification of water better and cheaper. Steenvoorden investigated if this method can become the standard method for the entire country. The unfortunate answer was no, due to policies of NGOs, bureaucracy in the governmental administration, and corruption.

Goal 7. Ensure access to affordable, reliable, sustainable and modern energy for all.
The paper on energy performance standards for houses (de Vries & Verhagen, 2016) reveals that energy savings are possible in a way that does not cost government anything and that stimulates innovation.

Goal 8. Promote sustained, inclusive and sustainable economic growth, full and productive employment and decent work for all.
Proper standardisation leads to better business and, as a consequence, to economic growth (Blind, 2004). The course addresses societal aspects as well and in that sense students are prepared to contribute to sustainability. If their future employers have the same vision remains to be seen, one of them is employed by a tobacco company …

Goal 9. Build resilient infrastructure, promote inclusive and sustainable industrialisation and foster innovation.
Of course, innovation is a core topic. Some examples address infrastructure—water management, rail roads, road pricing (to mitigate traffic jams and reduce emissions), and ICT infrastructure.

Goal 10. Reduce inequality within and among countries.
The comments to Goal 1 show the issue of inequality among countries. Inequality is also mentioned in relation to platforms and winner-takes-all mechanisms like network effects (Eisenmann et al., 2006) that may lead to extreme incomes for platform owners. The remedy is in alternative organisational forms such as cooperations, the feasibility of these is discussed.

Goal 11. Make cities and human settlements inclusive, safe, resilient and sustainable.
The role of standards in smart cities gets some attention including initiatives to improve safety in city neighbourhoods.

Goal 12. Ensure sustainable consumption and production patterns.
The company the students visit is Interface Europe Manufacturing. They produce carpet tiles in a sustainable way. In 2020 they want to be climate-neutral and then they intend to continue with 'climate take-back': a per saldo positive impact on the environment. In a group assignment students have to give advice to this company on how to continue. Additionally, the course mentions research on the business and

environmental impacts of the ISO 14001 standard for environmental management, the role of standards in life cycle analysis, and the impact of certification seals in influencing consumers in product perception and purchase intention.

Goal 13. Take urgent action to combat climate change and its impacts.
See the comments to Goals 7 and 12.

Goal 14. Conserve and sustainably use the oceans, seas and marine resources for sustainable development.
A Student Project Group in another course, the Minor Responsible Innovation, worked on the reduction of plastic supply to the sea via the Rotterdam harbour. They discovered that standards for sampling and measuring plastics are needed to be able to give evidence of pollution, trace sources of pollution, set priorities, and measure harbour cleaning performance afterwards (Vis, Johnson, & Stöver, 2016). A study on a standard for measuring poisonous substances in water caused by anti-fouling paint revealed that the standard was inaccurate, allowing paint industry to continue pollution (Gottlieb, Verheul, & de Vries, 2003). These examples are mentioned in discussing the importance of standards for measurements.

Goal 15. Protect, restore and promote sustainable use of terrestrial ecosystems, sustainably manage forests, combat desertification, and halt and reverse land degradation and halt biodiversity loss.
A project on standards and certification for sustainable tourism in the wetland area in the North of the Netherlands (Waddengebied) shows that standards and certification allow more tourism combined with lower impact on the ecosystem (Hickendorff, 2013). This study is mentioned as an example of the relevance of a local level of standardisation. A study on the Chain of Custody for sustainable wood shows how the current competing sets of requirements can be combined, allowing lower cost in the supply chain and a higher market share of sustainable wood, and thus less destruction of tropical rain forest (de Vries, Winter, & Willemse, 2017). This case is mentioned in another context, namely how role playing may help standardisation committees to achieve consensus.

Goal 16. Promote peaceful and inclusive societies for sustainable development, provide access to justice for all and build effective, accountable and inclusive institutions at all levels.
During the course, six different roles of government in relation to standardisation are discussed (de Vries, 1999), and the need for a national infrastructure for standardisation, innovation, metrology, quality management, certification and accreditation (Tippmann, 2013).

Goal 17. Strengthen the means of implementation and revitalise the global partnership for sustainable development.
Implementation of the SDGs in companies requires management systems. The International Organization for Standardization's 'High Level Structure' on management systems (Tangen & Warris, 2012) allows to integrate occupational health and safety

management, environmental management, energy management, integrity management, information security management or whatever area to be managed better, with quality management. It is mentioned in the course.

The SDGs are important for a sustainable future but can be criticized as well. In the first place, the current set of 17 is incomplete, the issues differ per case and also the stakeholders differ per case. In my course, students learn how to systematically trace stakeholders (per standard) and to distinguish core stakeholders from other stakeholders. These stakeholders have to agree, preferably consensus-based. Students learn how to achieve this in the context of standardisation committees, and how to combine design of a solution, and decision making in a way that the chances of market use of that solution are being enhanced.

The current goals relate to survival of mankind and healthy and happy living. Great, but there is more in life. Typically, the current goals reflect Western-humanistic values. Religions that dominate in other cultures are being ignored. Students get a glimpse of that in the paper by Haverkamp and de Vries (2016) that makes use of a Christian-philosophical approach to mitigate the tension between freedom and control, and in presenting the Islamic approach to business excellence developed by El Osrouti, de Vries, and Asif (2017, under review).

4 Discussion and Conclusion

From the perspective of the Endowed Chair on Standardisation the self-evident aim is to include standardisation in academic teaching. However, the elective course *Innovation and Interface Management* is part of the master programme *Management of Innovation*, and from that perspective the main aim should be to give students a better understanding of innovation management. Because standardisation is essential for innovation management, it is not difficult to combine these aims, but the second one should be leading. At School level, the third aim was added to also integrate sustainability in RSM's education, using the SDGs.

As Sect. 3 shows, the commonly acknowledged elements of sustainability management do get attention and the course even reaches beyond that. This does not imply that sustainability is treated in a systematic and balanced way. First because the course takes a managerial business perspective and thus not other lenses relevant for sustainability such as psychology, sociology or public policy. But also the managerial approach to sustainability is not developed in a way similar to the way interfaces and innovation are treated in this course, sustainability courses might be developed in a similar way. Starting from the sustainability side, both innovation and standardisation are essential, so both topics should be integrated in teaching programmes on sustainability. If defined in terms of the UN goals then the sustainability concept is multidimensional by nature, so research-based teaching requires a multidisciplinary approach. By the way, RSM even also offers an entire one-year master programme *Global Business and Sustainability*.

On the other hand, we have seen how a systematic approach of designing a standardisation curriculum automatically leads to the inclusion of sustainability. This is related to a combination of trends in the field of standardisation. Standardisation has a history of engineers and industrialists but meanwhile it has broadened in several ways:

- from products, components and technical processes to (also) services, management systems, and complex systems in which all of the previous ones are interconnected, e.g. smart cities;
- from business benefits to (also) societal benefits;
- from engineers, industrialists and (technical) scientists to a multitude of stakeholders, mainly private but also public.

As Sect. 3 shows, standardisation has become a core instrument for achieving and even for reaching beyond the SDGs. In innovation management, a similar trend can be observed: broadening towards *responsible* innovation (e.g., van den Hoven, Doorn, Swierstra, Koops, & Romijn, 2014). Innovation is no longer mainly 'inside out' but increasingly 'outside in' and then sustainability cannot be ignored, it can even form a starting point for innovation. And in any case, standards are indispensible. In this way, sustainability becomes a self-evident element of innovation management, and students who would never have chosen a specialisation in sustainability management get familiar with it.

The approach I chose for developing the *Innovation and Interface Management* elective at RSM is systematic but also based on my own experiences and research and the work done by my Ph.D. students and students. This makes it easier for me to teach but it is not necessarily best for someone else, probably not. This also questions the current attempts, among others by APEC, to come to common recognized standardisation curricula and related diplomas—it is probably too early for that. Apart from the lack of maturity of the research field and the huge diversity in business practices, also the differences in terms of university context hinder this—in my case it is a business school, and I integrate standardisation with innovation management whereas most colleagues teach it in a quality management context. At a technical university, a law school or a school of economics the emphasis should be different anyhow.

To conclude, this chapter shows how a standardisation curriculum can be designed in a systematic way and that then automatically also sustainability aspects will get addressed: most of the 17 UN Sustainable Development Goals get attention. This makes this chapter relevant not only for developing standardisation courses but also for integration of standardisation in sustainability courses.

References

Bateh, J., Horner, D. H., Broadbent, A., & Fish, D. (2014). Towards a theoretical integration of sustainability: A literature review and suggested way forward. *Journal of Sustainability Management, 2*(1), 35–42.

Bekkers, R. (2017). Where patents and standards come together. In R. Hawkins, K. Blind, & R. Page (Eds.), *Handbook of Standards and Innovation* (pp. 227–251). Cheltenham, UK / Northampton, MA: Edward Elgar Publishing.

Blind, K. (2004). *The economics of standards—Theory, evidence, policy.* Cheltenham, UK/Northampton, MA: Edward Elgar.

Bongers, C. (1980). *Standardization—Mathematical methods in assortment determination.* Boston, MA/The Hague, The Netherlands/London, UK: Martinus Nijhoff Publishing.

de Vries, H. (1997). Standardization—What's in a name? *Terminology—International Journal of Theoretical and Applied Issues in Specialized Communication, 4*(1), 55–83 (Rectification in 4(2)).

de Vries, H. (1998). The classification of standards. *Knowledge Organization, 25*(3), 79–89.

de Vries, H. J. (1999). *Standardization—A business approach to the role of national standardization organisations.* Boston, MA/Dordrecht, The Netherlands/London, UK: Kluwer Academic Publishers.

de Vries, H.J. (2005). Standardization education. *Homo Oeconomicus, 22*(1), 71–91.

de Vries, H. J. (2006). Best practice in company standardization. *International Journal for IT Standards and Standardization Research, 4*(1), 62–85.

de Vries, H. J. (2008). Standardisation: A business science perspective. In J. Schueler, A. Fickers, & A. Hommels (Eds.), *Bargaining norms, arguing standards—Negotiating technical standards, STT74* (pp. 18–32). The Hague: STT Netherlands Study Centre for Technology Trends.

de Vries, H. J. (2015). Standardisation—A developing field of research. In P. Delimatsis (Ed.) *The law, economics and politics of international standardisation.* Cambridge International Trade and Economics Law (pp. 19–41). Cambridge, UK: Cambridge University Press.

de Vries, H. J. (2016). *Het Vinkje verdwijnt ten onrechte.* Blog. Amsterdam: ESB. Retrieved January 2, 2018, from https://esb.nu/blog/20021103/het-vinkje-verdwijnt-ten-onrechte.

de Vries, H. J. (2018). *Product variety management* (Unpublished paper).

de Vries, H. J., de Ruijter, J. P. M., & Argam, N. (2011). Dominant design or multiple designs: The flash memory card case. *Technology Analysis & Strategic Management, 23*(3), 249–262.

de Vries, H. J., Feilzer, A. J., Gundlach, H. C. W., & Simons, C. A. J. (2010). Conformity assessment. In W. Hesser, A. J. Feilzer, & H. J. de Vries (Eds.), *Standardisation in companies and markets* (3rd ed., pp. 871–904). Hamburg, Germany: Helmut Schmidt University Hamburg.

de Vries, H. J., & Go, F. M. (2017). Developing a common standard for authentic restaurants. *The Service Industries Journal, 37*(15–16), 1008–1028.

de Vries, H. J., Nagtegaal, B., & Veenstra, S. (2017). Transatlantic harmonisation of standards and conformity assessment—A business perspective. In K. Jakobs & K. Blind (Eds.), *EURAS proceedings 2017—Digitisation: Challenge and opportunity for standardisation* (pp. T39–T58). Aachen, Germany: Mainz.

de Vries, H. J., & Verhagen, W. P. (2016). Impact of changes in regulatory performance standards on innovation: A case of energy performance standards for new-built houses. *Technovation, 48–49,* 56–68. https://doi.org/10.1016/j.technovation.2016.01.008.

de Vries, H. J., & Wiegmann, P. M. (2017). Impact of service standardization on service innovation. In R. Hawkins, K. Blind, & R. Page (Eds.), *Handbook of standards and innovation* (pp. 187–211). Cheltenham, UK: Edward Elgar.

de Vries, H. J., Winter, B., & Willemse, H. (2017). Achieving consensus despite apposing stakes: A case of national input for an ISO standard on sustainable wood. *International Journal of Standardization Research, 15*(1), 29–47.

den Uijl, S., Bekkers, R., & de Vries, H. J. (2013). Managing intellectual property using patent pools: Lessons from three generations of pools in the optical disc industry. *California Management Review, 55*(4), 31–50.

den Uijl, S., & de Vries, H. J. (2013). Pushing technological progress by strategic manoeuvring: The triumph of Blu-ray over HD-DVD. *Business History, 55*(8), 1361–1384. https://doi.org/10.1080/00076791.2013.771332.

Dul, J., Bruder, R., Buckle, P., Carayon, P., Falzon, P., Marras, W. S., et al. (2012). A strategy for human factors/ergonomics: Developing the discipline and profession. *Ergonomics, 55*(4), 377–395.

Dooyeweerd, H. (1955). *A new critique of theoretical thought 2: The general theory of the modal spheres.* Amsterdam, The Netherlands: Uitgeverij H. J. Paris/Philadelphia, PA: The Presbyterian Reformed Publishing Company.

Dooyeweerd, H. (1957). *A new critique of theoretical thought 3: The structures of individuality of temporal reality.* Amsterdam, The Netherlands: Uitgeverij H. J. Paris/Philadelphia, PA: The Presbyterian Reformed Publishing Company.

Dubois, A., & Gadde, L. (2002). The construction industry as a loosely coupled system: Implications for productivity and innovation. *Construction Management and Economics, 20*(10), 621–631.

Egyedi, T. M., & Ortt, J. R. (2017). Towards a functional classification of standards for innovation research. In R. Hawkins, K. Blind, & R. Page (Eds.) *Handbook of innovation and standards* (pp. 105–124). Cheltenham, UK/Northampton, MA: Edward Elgar.

Eisenmann, Th., Parker, G., & Van Alstyne, M. W. (2006). Strategies for two-sided markets. *Harvard Business Review, 84*(10), 92–101.

Elkington, J. (1999). *Cannibals with forks, the triple bottom line of the 21st century business.* Oxford, UK: Capstone Publishing.

El Osrouti, F., de Vries, H. J., & Asif, M. (2017). *Towards developing an Islamic business excellence model* (Unpublished paper).

Epstein, S. R., & Trak, M. (Eds.). (2008). *Guilds, innovation and the European economy, 1400–1800.* Cambridge, UK/New York, NY: Cambridge University Press.

Fickers, A., Hommels, A., & Schueler, J. (2008). Implications for research and policy. In A. Fickers, A. Hommels, & J. Schueler (Eds.), *Bargaining norms arguing standards, STT 74* (pp. 144–153). The Hague: STT Netherlands Study Centre for Technology Trends.

Gottlieb, A., Verheul, H., & de Vries, H. (2003). *Project Verbetering formele normalisatieproces—Case ISO 15181: Paints and varnishes—Determination of release rate of biocides in antifouling paints.* Delft/Rotterdam, The Netherlands: Centre for Process Management and Simulation.

Haverkamp, A., & de Vries, H. J. (2016). Managing in-company standardisation while avoiding resistance: A philosophical-empirical approach. In K. Jakobs (Ed.) *Effective standardization management in corporate settings.* Advances in IT standards and standardization research (pp. 184–213). Hershey, PA: IGI Global.

Hawkins, R., Blind, K., & Page, R. (Eds.). (2017). *Handbook of standards and innovation.* Cheltenham, UK: Edward Elgar.

Hickendorff, L. M. (2013). *Keurmerk voor duurzaam toerisme in Waddengebied—Draagvlak bij Texelse ondernemers* (Unpublished master thesis).

Karwowski, W. (Ed.). (2006). *Handbook of standards and guidelines in ergonomics and human factors.* Malwah, NJ: Lawrence Erlbaum Associates.

Langemeijer, M. M. (2004). *The Bangladesh shrimp chain: Playing or being played—A research paper on the consequences of international food safety standards on exports from developing countries* (Unpublished master thesis).

Lee, H. (2010). Standardization and innovation. In D.-G. Choi, B.-G. Kang, & T. Kim (Eds.), *Standardization: Fundamentals, impact, and business strategy* (pp. 153–177). Singapore: Asia-Pacific Economic Cooperation Secretariat.

Lee, H., Zoo, H., & de Vries, H. J. (2017). Interplay of innovation and standardization: Exploring the relevance in developing countries. *Technological Forecasting and Social Change, 118,* 334–348. https://doi.org/10.1016/j.techfore.2017.02.023.

Lelong, B., & Mallard, A. (Eds.). (2000). La fabrication des normes. In *Réseaux* (Vol. 18, No. 102, pp. 1–254). Paris: Hermes Science Publications.

Manders, B., de Vries, H. J., & Blind, K. (2016). ISO 9000 and product innovation: A literature review and conceptual model. *Technovation, 48–49*, 41–55. http://dx.doi.org/10.1016/j. technovation.2015.11.004.

Schumpeter, J. A. (1934). *The theory of economic development: An inquiry into profits, capital, credit, interest, and the business cycle.* New Brunswick, NJ/London, UK: Transaction Publishers.

Sherif, M. H. (2001). A framework for standardization in telecommunications and information technology. *IEEE Communications Magazine, 39*(4), 94–100.

Steenvoorden, G. P. (2008). *Buying behavior and processes in the Indonesian water sector—The micro hydraulic water treatment case* (Unpublished master thesis).

Tangen, S., & Warris, A.-M. (2012). Management makeover—New format for future ISO management system standards. *ISO Focus, 3*(7), 40–41.

Teichmann, H. (2010). *Global standardization: Coping with multilingualism. EURAS contributions to standardisation research 3.* Aachen, Germany: Mainz.

Tippmann, C. (2013). *National policies to attract R&D-intensive FDI in developing countries* (The Innovation Policy Platform. Policy Brief). Washington, DC, USA: The World Bank.

United Nations. (2018). *Sustainable development goals—17 goals to transform our world.* New York, NY: United Nations. Retrieved May 12, 2018, from https://www.un.org/ sustainabledevelopment/development-agenda/.

United Nations General Assembly. (2015). *Transforming our world: The 2030 agenda for sustainable development.* Resolution adopted by the General Assembly on 25 September 2015. A/RES/70/1. New York, NY: United Nations.

Université de Genève. (2018). *Standardization, social regulation and sustainable development.* Geneva: Université de Genève. Retrieved January 2, 2018, from http://www.standardization. unige.ch/.

Van de Kaa, G., van den Ende, J., de Vries, H. J., & van Heck, E. (2011). Factors for winning interface format battles: A review and synthesis of the literature. *Technological Forecasting & Social Change, 78*(8), 1397–1411. https://doi.org/10.1016/j.techfore.2011.03.011.

van de Lagemaat, D. (1986). *Onderwijzen in ondernemen.* Culemborg, The Netherlands: Educaboek.

van den Hoven, J., Doorn, N., Swierstra, T., Koops, B.-J., & Romijn, H. (Eds.). (2014). *Responsible innovation 1—Innovative solutions for global issues.* Dordrecht, The Netherlands: Springer.

Verman, L. C. (1973). *Standardization—A new discipline.* Hamden, CT: Archon Books, The Shoe String Press Inc.

Vis, A., Johnson, N., & Stöver, E. (2016). *Port waste catch* (Unpublished student report).

Wiegmann, P. M. (2016). Managing standard is in radical innovation projects—The case of micro combined heat and power. In K. Jakobs, A. Mione, A.-F. Cutting-Decelle, & S. Mignon (Eds.), *EURAS proceedings 2016—Co-opetition and open innovation* (pp. 455–472). Aachen, Germany: Mainz.

Wiegmann, P. M., de Vries, H. J., & Blind, K. (2017). Multi-mode standardisation: A critical review and a research agenda. *Research Policy, 46*(8), 1370–1386. https://doi.org/10.1016/j.respol.2017. 06.002.

World Commission on Environment and Development. (1987). *Our common future.* Oxford: Oxford University Press.

Henk J. de Vries (1957) is Professor of Standardisation Management at the Rotterdam School of Management, Erasmus University in Rotterdam, The Netherlands, Department of Technology and Operations Management, Section Innovation Management, and Visiting Professor at Delft University of Technology, Faculty of Technology, Policy and Management, Department of Values, Technology and Innovation, Section Economics of Technology and Innovation. His education and research focus on standardisation from a business point of view. From 1994 until 2003, Henk worked with NEN, Netherlands Standardization Institute, in several jobs, being responsible for R&D during the last period. Since 1994, he has an appointment at the Erasmus University's School of Management and since 2004, he has been working full-time at this university.

Henk is (co-)author of more than 380 publications on standardisation, including several books. See www.rsm.nl/people/henk-de-vries. In 2009, the International Organization for Standardization ISO awarded his education about standardisation as best in the world. Henk is President of the European Academy for Standardisation EURAS.

Cases of Countries

Education About Standardization in the Context of Sustainable Development

Elka Vasileva

1 Introduction

The economic and social impact of standardization and standards on sustainable development in the 21st century is indisputable. This justifies the growing need for education about standardization for the implementation of contemporary models for sustainable consumption and production. In this respect the development of standardization in the former socialist countries of the so-called Soviet bloc in Central and Eastern Europe provides opportunities for the chronological study of these issues. In the transition from the socialist model of governance to the current market economy, standardization and standards have changed their role—from a *legal regulator of the centrally planned economy* into a *"new form of regulation* in the modern, globalized life alongside traditional legislation and normative society" (Bostrom & Klintman, 2011; Brunsson & Jacobson, 2000; Busch, 2000; Rasche, 2012, 2015; Waddock, 2008). An important element of this transition relates to the European standardization policy that needs to be addressed as part of the process of accession of the countries of Central and Eastern Europe to the European Union. Therefore, Bulgarian standardization had to be brought into line with the modern principles for carrying out this activity. We will use this country currently member of the European Union to study this.

Of course, these changes have implications for education about standardization. This contribution aims to examine education about standardization chronologically to the development of Bulgaria: from the socialist era, during the transition period to the present day market economy, at the beginning of the 21st century, with the necessary discourses on sustainable development. This article aims also to study the role of education about standardization in the framework of university education in the country.

E. Vasileva (✉)
University of National and World Economy, Sofia, Bulgaria
e-mail: elkav@unwe.bg

© Springer Nature Switzerland AG 2020
S. O. Idowu et al. (eds.), *Sustainable Development*, CSR, Sustainability,
Ethics & Governance, https://doi.org/10.1007/978-3-030-28715-3_6

The presented chapter is structured in three parts. In Sect. 2, education about standardization during the different stages of development of Bulgaria is traced: from the socialist era and the mandatory standards (Sect. 2.1) to the transition phase to the market economy and the role of the European standards (Sect. 2.2). Section 3 examines the contemporary phase of standardization and market economy. Section 4 provides a more detailed specific example of academic education about standardization in the field of economics and management. The *Concluding section* contains a summary of the key findings.

2 Chronology of Education About Standardization in Bulgaria

This section presents a diachronic development of Bulgaria, with the respective discourses on education about standardization—from the socialist era through the period of transition to the present day market economy.

2.1 Socialist Era and Mandatory Standards

Bulgaria, like other countries of Central and Eastern Europe, passed through a complex and full of challenges path of changes after the fall of the "iron curtain" in 1989. The political, social and economic changes that followed can be tracked and analysed only if the socialist model of governance, "dramatically different from that of the Western countries" (Hoogstoel, 1998), is considered. The introduction of state-owned property in the era of socialism predetermined the highly politicized management of state-owned enterprises. There was little or no private ownership of property. The state controlled these enterprises through its central planning system and the implementation of so-called five-year plans. Thus directors became executors of decisions taken by the top ranks of the political hierarchy, often pressured to "meet the plan", even taking shortcuts to meet the plan (Hoogstoel, 1998; Iankova, 2009). Their main function was to achieve the planned goal rather than to take appropriate management decisions. On the other hand, the state was the only customer, thus the possibility of effective feedback from users and their involvement in the development of the goods and services was ignored. Numerous specially created central agencies represented the needs and demands of the population concerning consumer goods. In reality, in the context of deficit, consumers bought the goods which were available on the market within a very limited range, and quality suffered (Ghodsee, 2007; Egermayer, 1988; Koleva, 2006).

A system for state quality assessment of goods was developed, according to which three categories or grades were awarded to the production: grade "K" for products with quality above the world average, grade "1" for products with the quality of the world average and grade "2" for products with quality below the world average (Ribov & Andreev, 1990; Hoogstoel, 1998). The "highest" and "first" grade products generally carried an approval stamp of certification from a national laboratory.

For the purpose of state assessment and quality control of goods, Bulgarian State Standards (BDS) were developed, which were validated at national level and were mandatory for the whole country. For 1990, Ribov and Andreev (1990) quote that around 13,000 BDSs were in force in the country and more than 20 "*Uniform systems of standards for a given facility or area*" were developed (Ribov & Andreev, 1990, pp. 51). The full standardization activity in Bulgaria was carried out under the guidance of the Quality Committee of the Council of Ministers, by the "Standardization" department especially created for this purpose.

Most of the Central and Eastern European countries, led by the Soviet Union, formed a trade union called Council for Mutual Economic Assistance (CMEA). Between 1971 and 1980, the development and adoption of a wide variety of uniform quality standards (CMEA standards) was seen as an instrument to promote economic integration among member states. The process of "harmonizing the national standards of the member states in the framework of the socialist economic integration" (Ribov & Andreev, 1990, pp. 58) led to the possibility of the standards of the CMEA to be applied *directly* as Bulgarian state standards, without any alteration.

A study focused on the food industry in the country's socialist period cites evidence that "there is a strict quality control regulation," and "the people employed in the industry without exception find the control for compliance with Bulgarian state standards, both concerning hygiene and sanitary requirements, as extremely strict". The author draws attention to the fact that, in practice, numerous deviations from the rules and standards are allowed as one of the "elements of the reality of planned economy" (Shkodrova, 2014, pp. 18 and pp. 49). On the other hand, Egermayer (1988) emphasizes that the implementation of the system of state assessment and quality control of goods in the CMEA countries is "not uniform, and the results of the award of categories of quality assessments are not internationally valid".

In order to support the functioning of this standardization system, education about standardization was widespread, both in secondary schools (in specialized vocational schools and high schools), as well as in higher education institutions. Subjects concerning so-called "Technical Standardization" were taught at higher technical schools, while in economic higher schools education was focused on "Standardization and Quality Management" and on standards for products, raw materials and commodities within the discipline "Commodity Science and Technology". These academic courses covered topics related to: *the national standardization system* (development of standardization documents, classification, categories and types of standards, principles of standardization, etc.); *international and regional standardization* (ISO and CEN (European Committee for Standardization) and *standardization within the framework of the CMEA* (uniform standards of quality or standards of the CMEA). The study of standards was also included in the topics of assessment, control and quality management of goods. The great number of classes in the courses, as well as the practical orientation of the seminars on these topics is impressive.

2.2 The Period of Transition to a Market Economy—The Role of European Standards

After the collapse of the Socialist Bloc and the fall of the Berlin Wall in 1989, the countries of Central and Eastern Europe began a transition to a free market economy from different starting points. For Bulgaria, this transitional phase was characterized by a sharp decline in the GDP and employment, enormous losses of human and physical capital, erosion in supplier-consumer relations, and loss of foreign markets.

In the period 1995–2007, the economies of Central and Eastern European countries began to emerge from the economic collapse. Bulgaria also entered a period of catching up and orientation towards accession to the European Union. The countries of Central and Eastern Europe went through this phase of transition in different ways. Unlike the countries which joined the EU in 2004, Bulgaria was lagging behind in economic restructuring and transformation and, as a consequence, in the implementation of EU accession economic conditions. Iankova (2009) concludes that Bulgaria differed from the other candidates for EU membership "in the individualism of its post-communist transformation and its insufficient readiness to join the EU" (Iankova, 2009, pp. 6).

In the sphere of standardization, serious transformations were also taking place, and in 1999 the legal framework was changed with the adoption of a new National Standardization Act (SG, issue 55 of June 18, 1999). It introduced the principle of *voluntary action* in line with the principles of European standardization. The status of the Bulgarian Institute for Standardization (BIS), the newly established Bulgarian standardization body, changed from an administrative body of the executive power to a public-law organization (Burov, 2017). Pre-accession efforts led to the introduction of European standardization requirements in the work of both the public authority and all private, public and societal stakeholders.

In 2003 alone, 5576 European Standards (EN) were introduced as Bulgarian Standards (BDS) (BIS, 2007, pp. 18). In 2006—the year preceding the entry of Bulgaria into the EU—the total number of European standards introduced as BDS was 4342 (Fig. 1). The larger number of adopted European standards in 2003 and 2006 was related to the fulfilment of the criteria for BIS full membership in the European Standards Organizations CEN and CENELEC. All new CEN/CENELEC standards were adopted as BDS national standards in 2007.

In the process of Bulgaria's accession to the European Union there were also problems that impeded the practical application of the EU standards and those of the "CE" marking system (European Conformity Marking). They were mainly related to the fact that the greater part of the EU standards, adopted as BDS, were introduced in English. The situation that the standards are not translated into Bulgarian creates problems in their future use in the country. This lack of translations led to the use of incorrect terms in a given area of standardization and affected the quality infrastructure (e.g. the activities of the testing and calibration laboratories, etc.).

The education about standardization sharply shrank, with most of the curricula both in secondary schools and higher education institutions disappearing or signifi-

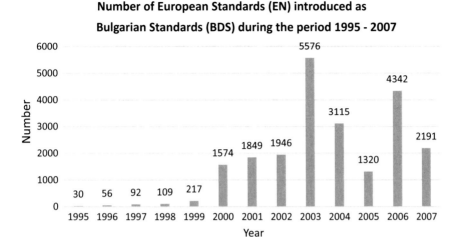

Fig. 1 Number of European Standards (EN) introduced as Bulgarian Standards (BDS), according to BIS sources (BIS, 2007)

cantly declining. This was mainly due to the structural changes in education in the country and to the lack of public understanding of the role of the European standards for the economy and society. Standardization education in this period was linked both to content describing the socialist system of standardization and to new content reflecting the changes that had occurred in the light of the European standardization.

The education about standardization at technical higher education institutions was oriented towards standardization in support of the European policies and legislation of the "New Approach" and the CE marking system.

Similar to the trend in many European economic university members of International Society of Commodity Science and Technology (IGWT) in Bulgaria too the education about standardization in the curricula of "Commodity Science" was increasingly neglected and even removed from the curriculum (Waginger, 2010; Wolfgang, 2004).

Undoubtedly, standardization education in universities in the countries of the former Soviet Bloc had been influenced by past traditions when standardization was an essential part of the central economy. This acted as a hindering factor in the acceptance and appreciation of voluntary standardization during the transition period. The tendencies of limiting the topics related to standardization in the curriculum of "Commodity Science" were confirmed by the numerous discussions which took place during the International Symposiums of IGWT. It is interesting to note that the lack of willingness of universities to continue teaching in the area of standardization affected both the countries of the Central and Eastern Europe (Bulgaria, Lithuania, Poland, Romania, Slovenia, Slovakia, Ukraine, etc.) and the Western member countries of the IGWT such as Italy, Germany and Austria (Waginger, 2010; Wolfgang, 2004). At the same time, efforts were made to cooperate among IGWT members to create a common syllabus of "Standardization", reflect the impact of the European standardization system (IGWT, 2010, 2014, 2016).

3 Contemporary Development of Education About Standardization in the Market Economy

The contemporary phase of standardization development in Bulgaria is related to the impact of the European policy in the sphere of standardization in the period after the country's accession to the EU in 2007.

3.1 The Nostalgia for the "Mandatory Standards"

In recent years, the nostalgia for the "mandatory standards" on the Bulgarian market has been manifested in the country, in particular in the area of food quality. In order to preserve the unique taste and the national specific characteristics of certain foods on the market, initiatives have been taken in the country over the last ten years to update and develop national and industry standards. They come in response to the increased public needs of objective quality information, which needs arise, from the highly compromised traditions in the production of Bulgarian dairy and meat products in the transition period. There is a tendency for today's Bulgarian consumers to seek quality corresponding to the "old" BDS food standards with mandatory requirements from the time of the planned socialist economy (Vasileva & Ivanova, 2012).

A number of initiatives have been launched to create a labelling differentiation of the products made in accordance with the BDS requirements for emblematic foods for Bulgarian consumers—Bulgarian yoghurt and other dairy products, bread, meat specialties, etc. For their analysis a series of criteria, applicable to the assessment of third party certification schemes, were used in Bulgaria (Nikolov & Vasileva, 2010; Nikolov, Vassileva, & Ivanova, 2010): *scope/level* (products covered by the label and the level of application); *organizers/owners*; *stakeholders* (needs satisfaction, dialogue); application of the principles of *impartiality, accessibility, independence and transparency*; *providing information to consumers*; *product specifications* (BDS standards, environmental aspects, etc.); *objective assessment and approval* (compliance with criteria and/or standards); *presentation* (typical logo, certificate, etc.).

Researches by Vasileva and Ivanova (2012) revealed serious deficiencies in the knowledge about the definitions, principles and goals of standardization, types of standards, and the stages of the development of national BDS standards (Nikolov, Vassileva, & Ivanova, 2010). The lack of application of the standardization principles of impartiality, accessibility, independence and transparency (Nikolov & Vasileva, 2010) was found established in the case of the Stara Planina label (the name of the

Balkan mountain in Bulgarian). A significant knowledge deficit on standards and standardization affecting all stakeholders along the value chain (consumers, manufacturers, traders, government institutions, education institutions, sectoral and consumer organizations etc.) has been confirmed (Nikolov & Vasileva, 2010; Nikolov, Vassileva, & Ivanova, 2010). This calls for a wide-ranging discussion on the dissemination of standardization knowledge among all those interested in food quality on the Bulgarian market. A great potential in this direction is the training at the different phases of education including in academic institutions.

3.2 The Bulgarian Institute of Standardization Cooperates with the Academic Institutions

As part of its activities the Bulgarian Institute of Standardization has expanded its cooperation with the academic institutions in the country. The Institute nominated education programs on standardization to participate in the ISO Award "Higher Education in Standardization" (ISO, 2009) competition, and thus motivated academic circles to include standardization in higher education institutions in Bulgaria.

On the basis of an overview of the comparatively limited number of academic programs in 2009, the Bulgarian Institute of Standardization awarded the curriculum of "Standardization", developed by the Chair of Commodity Science, University of National and World Economy—Sofia. The curriculum provides knowledge and skills about the role and use of standardization in relation to technologies, commerce and quality management. Although the program did not win the ISO Award, this was an opportunity for benchmarking in the education about standardization at national level in the transition period (see below "Bulgarian case study").

In recent years, the Bulgarian Institute for Standardization, in cooperation with higher education institutions, has developed a series of initiatives to promote standards among Bulgarian students. As a result of this cooperation, *standardization information centres* have been opened, and in 2016 five new centres were opened in higher education institutions throughout the country (BIS, 2016, pp. 12). In these information hubs, students and teachers are provided with free access to reading materials on standards from different fields, depending on the specifics of the higher education institution.

3.3 Education About Standardization at the Universities of Bulgaria

Despite the efforts of the Bulgarian national standardization body to promote its activities and the role of standards, education about standardization within the framework of university education remains unsystematic and limited. Education about standardization is conducted within the bachelor's or master's degrees programmes. In most of them, the study of standards is included as a whole syllabus or as a separate subject in a syllabus.

The review of the current situation of academic education about standardization in Bulgaria is based on a few numbers of studies as objective sources of information. They mainly concern education in the field of economics and management, and this hampers the assumptions at national level.

In 2012, the "Corporate Social Responsibility (CSR) and Sustainable Development" survey showed that in the curricula of Bulgaria's 51 accredited schools of higher education, 19 universities (38%) include some of the Corporate Social Responsibility (CSR) and sustainable development standards related issues in their bachelor's or master's degrees programmes (Simeonov & Stefanova, 2015, pp. 328; Slavova & Bankova, 2015; Horváth & Pütter, 2017).

Slavova and Bankova's survey focuses on the inclusion of the broad CSR topic in business and management education in higher education institutions in Bulgaria (Slavova & Bankova, 2015). The authors use web-content analysis of the curricula and study programs of all Bulgarian universities training business and management students in bachelor's and master's programmes. The results show that more than 80% of the universities in Bulgaria have included CSR-related disciplines (ethics, CSR, sustainable development, environmental management) as standalone courses in their curricula. The study does not specify whether specific CSR standards are being studied in the enumerated disciplines. This Bulgarian study agrees with other authors that CSR education as a standalone discipline runs the risk of being viewed as a separate issue, isolated and unrelated to core business courses (Matten & Moon, 2004).

The survey points out that disciplines related to environmental problems (Environmental Management; Ecology and Environmental Protection) and sustainable development are the ones most often offered in universities (approximately 50% of all disciplines from the sample (Slavova & Bankova, 2015). These facts the authors connect with the requirements of the business environment in the country: the existing environmental legislation; the introduction and practice of environmental impact assessment; the growing number of companies which have obtained ISO 14001 certification, the corporate social activities carried out by business organizations related to achieving an economic and environmental impact, etc. In this indirect way, the availability of topics related to the ISO 14000 series of standards in business and management education in higher education institutions in Bulgaria can be confirmed.

The findings of the authors of this study support the thesis that CSR (CSR standards included) should be embedded in the curriculum at different levels of education because "mainstreaming" teaching provides opportunities for every business student to be informed about the social and ethical dimensions in their future work as a business person (Slavova and Bankova, 2015).

A survey among Bulgarian higher education institutions, made in 2014 on the occasion of the establishment of the Bulgarian Academic Quality Network (Vasileva, 2014), found that the *quality management standards of the ISO 9000 series* are widely taught in higher education institutions in Bulgaria. Progress is also being made in education about the standards for *environmental management and social responsibility* (Table 1).

Twenty-five of all higher education institutions in Bulgaria which provide Business and Management education to students, according to the National Register of the Ministry of Education and Science of the Republic of Bulgaria and the methodology of Slavova and Bankova are presented in Table 1. These are the universities that have received accreditation from the National Evaluation and Accreditation Agency in the professional fields "Administration and Management, and Economics" for bachelor's and master's degree.

Information on sustainability standards-related courses has been gathered from interviews with representatives of the administration of the surveyed universities conducted within the Bulgarian Academic Quality Network (Vasileva, 2014). Only the presence or absence of such courses is shown in Table 1. Standards teaching are examined in three main sustainability standards groups: *certification standards* of organizations management systems (ISO 9001 and ISO 14001); *labelling standards* of product and services (eco-labelling standards, LCA standards etc.) and *CSR Guidelines* (ISO 26000).

Education related to the standards for quality management and environmental management systems (ISO 9001 and ISO 14001) is dominant. This is due to the growing interest of companies in the country to certify management systems according to the requirements of ISO 9001 and ISO 14001 and the expectations of the business in relation to the quality management and environmental management skills and competences.

Table 1 shows that topics related to labelling standards of product and services are scarcely present in academic education. Studying the environmental aspects of products and services and LCA standards, requires a wider presence in the curricula. Again, we can say that the business environment in the country does not foster an interest in such education. Current research shows that Bulgarian companies are not aware of the benefits of ISO 14001 standards as well as of the European Eco-Management and Audit Scheme (EMAS) and EU Ecolabel (Slavova, 2015; Ivanova, Haradinova, & Vasileva 2016). Practice shows that in Bulgaria only three organizations are registered under EMAS and only one company fulfills the EU Ecolabel criteria.

A relatively small number of universities state they offer courses on CSR Guidelines standards, including ISO 26000, which corresponds to other Bulgarian studies (Simeonov & Stefanova, 2015; Slavova & Bankova, 2015). This can be explained by the fact that CSR is a relatively new concept and it has become more and more widely spread in practice recently. Various forms of corporate social initiatives have been identified and a number of companies which are subsidiaries of multinational companies stand out with an integrated approach to the implementation of CSR programs and projects (Slavova, 2015). Unlike other Western European countries global companies operating on the Bulgarian market, often supported by international institutions such as the UN Development Program, have a defining role in the emergence of CSR initiatives in Bulgaria.

Table 1 Sustainability standards-related courses included in the business and management education in Bulgarian universities

No	Universities	Courses		
		1 Management systems	2 Labelling	3 CSR
1.	Agricultural University		X	
2.	Angel Kanchev University of Ruse	X		
3.	Burgas Free University			
4.	International Business School	X		
5.	New Bulgarian University	X		X
6.	Plovdiv University "Paisii Hilendarski"	X		
7.	Prof. Assen Zlatarov University	X		
8.	Sofia University "St. Kliment Ohridski"	X		X
9.	South-West University "Neofit Rilski"	X		X
10.	St. Cyril and St. Methodius University—Veliko Tarnovo	X		
11.	Technical University—Gabrovo	X		
12.	Technical University of Sofia	X		
13.	Technical University-Varna	X		
14.	Todor Kableshkov University of Transport	X		
15.	Trakia University—Stara Zagora	X	X	
16.	Tsenov Academy of Economics	X		X
17.	University of Agribusiness and Rural Development			
18.	University of Chemical Technology and Metallurgy	X		
19.	University of Economics—Varna	X	X	X
20.	University of Finance, Business and Entrepreneurship	X		X
21.	University of Food Technologies	X	X	
22.	University of Forestry	X		
23.	University of National and World Economy	X	X	X
24.	Varna Free University "Chernorizets Hrabar"	X		
25.	Vasil Levski National Military University	X		

Source Survey within Bulgarian Academic Quality Network (Vasileva, 2014)
Notes 1—Certification standards of organizations management systems (ISO 9001 and ISO 14001); 2—Labelling standards of product and services (eco-labelling standards, LCA standards); 3—CSR Guidelines (ISO 26000)

Obviously, Bulgarian universities have to respond to the global sustainable development challenges related to sustainable development and incorporate sustainability standards-related topics into curricula in the field of Administration and Management and Economics, in line with the findings of other authors (Heidt & Lamberton, 2011; Matten & Moon, 2004).

This can be done in the form of the standalone sustainability standards disciplines, according to the proposed typology of standards in Table 1. This will allow education about standardization to be integrated into other curricula that address wider issues of sustainable development (Stubbs & Cocklin, 2008).

The national context will also influence the trends in education about sustainability standards and the scope of this education in Bulgarian universities will expand with the introduction of the concept of sustainable development.

4 Education About Standardization—The Case of Bulgaria

This part describes in more detail a specific example of academic education about standardization in the field of economics and management. To that end, the *framework* for standardization education, developed by Choi and de Vries (2011) is used because it provides a coherently structured and unified set of five components: target groups (who); appropriate learning objectives (why); probable programme operators (where); prospective content modules (what) and teaching methods (how) (Choi & de Vries, 2011).

Our case is the "Standardization" programme that has been developed by the Chair of Commodity Science, University of National and World Economy—Sofia. In 2009 it was nominated by the Bulgarian Institute of Standardisation for participation in the ISO Award "Higher Education in Standardization" (ISO, 2009). At that time, this curriculum provided the basis for teaching standardization according to the principles of the European standardization i.e. it overturned the concept of teaching standards in Bulgarian context. Despite its "classic" look from the contemporary point of view, this small standardization module represents a good practice in the Bulgarian academic context to provide the basic knowledge and skills in the field of standardization.

4.1 Target Groups (Who)

The academic programme has been developed for the "Master" qualification and education degree and has been included in the curriculum of the specialties "Economic of Logistics" and "Economics of Transport" at the University of National and World Economy, Sofia, Bulgaria. The programme is aimed at students from several nationalities—Bulgarian, Greek, and Macedonian. Four credits are awarded for this programme, equal to 112 h.

4.2 Appropriate Learning Objectives (Why)

The objectives of the programme are to give the following:

- *Knowledge*—by studying "Standardization" students gain basic knowledge of the essence, the subject, the aims, the tasks and the basic principles of standardization; knowledge of the structure of the global (international) system of standardisation (ISO), the regional and national systems of standardization; knowledge of the principles of conformity assessment (including CE marking), the structure and operation of the certification systems; knowledge of the role and the function of the accreditation bodies and market surveillance;
- *Skills*—the course provides opportunities for acquiring skills of practical work with standards, and certification of products and services. The course emphasises on the issues relating to quality control and assurance with the tools which standardization and certification provide.

4.3 Programme Operators (Where)

The initiative for the development and education in the programme is introduced by a team of lecturers at the Department of Commodity Science at the University of National and World Economy—Sofia. This initiative is proposed to be extended within the education in commodity studies field with the IGWT university community (Austria, Poland, Lithuania, Slovenia, Ukraine, and Romania).

4.4 Programme Contents Modules (What)

The academic programme "Standardization" studies: the essence, the subject, the aims, tasks and the basic principles of *standardization*; the structure of the international, regional and national system of Standardization; the basic procedures in the development and updating of standards, and their classification. In close relation to the knowledge of the problems of standardization is the necessity to acquire understanding about the essence and importance of *certification*, the types of certification and the certification systems, the key stages and procedures in certification. The course emphases the issues relating to *quality control and assurance* with the tools of standardization and certification as well as the role and the function of *accreditation* bodies and *market surveillance*.

4.5 Teaching Methods (How)

The programme envisages a course of lectures (30 academic classes) and seminars (30 academic classes). The seminars include discussion on preliminary set topics, individual and group assignments with the purpose of expanding the gained knowledge and acquisition of skills of practical application of the knowledge of standardization, certification, quality control and assurance. During these seminar classes, students solve real-life *case studies* and *discuss specific problems*. Students with similar interests work together in small teams where they exchange experiences and solve group assignments on practical problems.

The Department has classrooms and specialized laboratories, with equipment for practical quality testing by application of standardized methods. Evaluation of the knowledge acquired in lectures is done in seminar classes during the semester. Based on class participation, attendance and the results shown in individual and group assignments students are awarded grades for seminar classes (according to a six points grading scale).

The presented academic programme "Standardization" has been developed for the purpose of education in economics and management students in "Master" qualification and educational degree. It provides *knowledge and skills* about the role and use of standardization in relation to logistics, transport, commerce and quality control and assurance. The main findings of Choi and de Vries (2011) for the existence of the so-called Module 3 [Academic/Theoretical aspects] and Module 4 [Case studies] in the Higher-Education Orientation Domain (Choi & de Vries, 2011; de Vries, 2002; Olshefsky, 2008) are confirmed by the *programme content* described above. The listed methods applied in seminar classes are also in line with the content identified by the authors in Module 4 [Case studies], case studies, group activities, and hands-on experiences as teaching modalities.

The "Standardization" academic programme presented in this part of the article can also be commented upon in the development of standardization-related competencies prospective (Blind & Drechsler, 2017). Parallel to the *general competencies (to know basic terms in the fields of standardisation; to know the international committee landscape; to be able to identify different types of standards; too be able to research standards; to be able to screen a standards; to be able to apply standardized methods of quality control and assurance)* provided by the higher education institutions in the field of standardization, this programme also enables the creation of a number of *"soft" skills (interpersonal communication; flexibility, team work; time management, presentation skills* etc.) mentioned in Blind and Drechsler's hierarchical competence model (Blind & Drechsler, 2017).

5 Conclusion

In this chapter, education about standardization was examined at the different stages of development in Bulgaria: from *the socialist era* and the mandatory standards through *the transitional phase* to the market economy and the role of European standards to *the contemporary stage* of standardization and the market economy. The unique manifestation of nostalgia for the "mandatory standards" of food quality on the Bulgarian market shows the significant knowledge deficit on standards and standardization affecting all stakeholders along the value chain (consumers, manufacturers, traders, state institutions, educational institutions, sectoral and consumer organizations, etc.). In this context, current national practices for academic education about standardization have been presented, and a more detailed specific example of academic education about standardization in the field of economics and management.

In conclusion it may be noted that the current state of education about standardization in Bulgaria is highly fragmented. However, the topics of the role of standardization in the world of globalization and the efforts for sustainable development are becoming more important in the context of Bulgaria too, and as a result get increased attention in university programmes.

References

BIS (Bulgarian Institute for Standardization). (2007). *Annual Report 2007*. Sofia, Bulgaria: Standartizacia print Ltd.

BIS (Bulgarian Institute for Standardization). (2016). *Annual Report 2016*. Sofia, Bulgaria: Standartizacia print Ltd.

Blind, K., & Drechsler, S. (2017). *European market needs for education in standardisation/standardisation-related competence, European commission*. Directorate-General for Internal Market, Industry, Entrepreneurship and SMEs, Unit B3 Standards for Growth. Luxembourg: Publications Office of the European Union.

Bostrom, M., & Klintman, M. (2011). *Eco-standards, product labelling and green consumerism*. Hampshire, UK: Palgrave Macmillan.

Brunsson, N., & Jacobsson, B. (2000). The contemporary expansion of standardization. In N. Brunsson & B. Jacobsson (Eds.), *A world of standards* (pp. 1–17). Oxford, UK and New York, USA: Oxford University Press.

Burov, I. (2017). *10 years of the BIS as a public-law organization* (Bulgarian institute for standardization. Annual Report 2016). Sofia, Bulgaria: Standartizacia print Ltd.

Busch, L. (2000). The moral economy of grade and standards. *Journal of Rural Studies, 16*, 273–283. https://doi.org/10.1016/S0743-0167(99)00061-3.

Choi, D. G., & de Vries, H. J. (2011). Standardization as emerging content in technology education at all levels of education. *International Journal of Technology and Design Education, 21*, 111–135. https://doi.org/10.1007/s10798-009-9110-z.

de Vries, H. J. (2002). *Standardisation education* (ERS-2002-82-ORG). Rotterdam: ERIM report series "Research in management".

Egermayer, F. (1988). Quality in socialist countries. In J. M. Juran (Ed.), *Quality control handbook* (pp. 35–40). New York, USA: McGraw-Hill.

Education About Standardization in the Context of Sustainable … 93

Ghodsee, K. (2007). Potions, lotions and lipstick: The gendered consumption of cosmetics and perfumery in socialist and post-socialist urban Bulgaria. *Women's Studies International Forum, 30,* 26–39. https://doi.org/10.1016/j.wsif.2006.12.003.

Heidt, T., & Lamberton, G. (2011). Sustainability in the undergraduate and postgraduate business curriculum of a regional university: A critical perspective. *Journal of Management & Organization, 17*(5), 670–690. https://doi.org/10.1017/S1833367200001322.

Hoogstoel, R. E. (1998). Quality in central and eastern Europe. In J. M. Juran & A. Blanton Godfrey (Eds.), *Juran's quality handbook* (pp. 39–41). New York, USA: McGraw-Hill.

Horváth, P., & Pütter, J. M. (2017). Sustainability reporting in central and Eastern European companies. International Empirical Insights. In P. Horváth & J. M. Pütter (Eds.), *MIR series in international business.* Springer International Publishing.

Iankova, E. (2009). *Business, government, and EU accession strategic partnership and conflict.* Lexington, UK: Lexington Book.

IGWT. *International Society of Commodity Science and Technology, International Symposiums.* http://www.igwt.uek.krakow.pl/.

IGWT. (2010, September). *17th IGWT Symposium, "Facing the Challenges of the Future: Excellence in Business and Commodity Science",* Bucharest, Romania.

IGWT. (2014, September). *19th IGWT Symposium "Commodity Science in Research and Practice—Current Achievements and Future Challenges",* Crakow, Poland.

IGWT. (2016, September). *20th IGWT Symposium, "Commodity Science in a Changing World",* Varna, Bulgaria.

ISO (2009) ISO Award "Higher Education in Standardization". ISO, Geneva, Switzerland. Available online from https://www.iso.org/files/live/sites/isoorg/files/archive/pdf/en/iso_award_2009.pdf. Accessed 06.07.2018.

Ivanova, D., Haradinova, A., & Vasileva, E. (2016). Environmental performance of companies with environmental management systems in Bulgaria. *Quality—Access to Success Journal, 152*(17), 61–66.

Koleva, P. (2006). *Nouvelles Europes, Trajectoires et enjeux économiques.* Belfort-Montbéliard, France: Université de Technologie.

Matten, D., & Moon, J. (2004). Corporate social responsibility education in Europe. *Journal of Business Ethics, 54,* 323–337. https://doi.org/10.1023/B:BUSI.0000049886.47295.3b.

National Standardization Act, State Gazette, issue 55 of June 18, 1999.

Nikolov, B., & Vasileva, E. (2010, November). Quality information through food labeling. The "Stara Planina" case. In *XXI National Scientific and Practical Conference "Quality—For a Better Life",* Sofia, Bulgaria.

Nikolov, B., Vassileva, E., & Ivanova, D. (2010). Quality labelling and information asymmetry: Identification of quality labelling certification schemes in Bulgaria. In M. Zoeva (Eds.), Quality—Access to Success Journal: Volume I. *Excellence in business and commodity science* (pp. 83–92).

Olshefsky, J. (2008). ASTM international—The strategic value of standards education. In D. E. Purcell (Ed.), *The strategic value of standards education—A global survey* (pp. 5–17). Washington D.C., USA: The Center for Global Standards Analysis.

Rasche, A. (2012). Global policies and local practice: Loose and tight couplings in multi- stakeholder initiatives. *Business Ethics Quarterly, 22*(4), 679–708. https://doi.org/10.5840/beq201222444.

Rasche, A. (2015). Voluntary standards as enablers and impediments to sustainable consumption. In L. Reisch & J. Thøgersen (Eds.), *Handbook of research on sustainable consumption* (pp. 343–358). Cheltenham, UK: Edward Elgar Publishing.

Ribov, M., & Andreev, A. (1990). *Standardization and quality management.* Sofia, Bulgaria: Higher Economic Institute "Karl Marx".

Simeonov, S., & Stefanova, M. (2015). Corporate social responsibility in Bulgaria: The current state of the field. In S. Idowu, R. Schmidpeter, & M. Fifka (Eds.), *Corporate social responsibility in Europe. CSR, sustainability, ethics & governance* (pp. 313–332). Cham: Springer.

Shkodrova, A. (2014). *Communist gourmet. The curious history of food in People's Republic of Bulgaria.* Plovdiv, Bulgaria: Janet 45.

Slavova, I., & Bankova, Y. (2015). Corporate social responsibility in business and management university education: The relevancy to the business practices in Bulgaria. *European Journal of Business and Economics, 10*(2), 57–64. https://doi.org/10.12955/ejbe.v10i2.697.

Slavova, I. (2015). Corporate social responsibility in Bulgaria: Development, constraints and challenges. *Economic and Social Alternatives, 2,* 113–126.

Stubbs, W., & Cocklin, C. (2008). Teaching sustainability to business students: Shifting mindsets. *International Journal of Sustainability in Higher Education, 9*(3), 206–221. https://doi.org/10.1108/14676370810885844.

Vasileva, E. (2014, November). Bulgarian academic quality network. In *25-th National Scientific and Practical Conference "Quality—For a Better Life"*, Sofia, Bulgaria.

Vasileva, E., & Ivanova, D. (2012). Consumer behaviour and food consumption patterns in south east Europe. In D. Sternad & T. Döring (Eds.), *Handbook of doing business in south east Europe* (pp. 271–293). Hampshire, UK: Palgrave Macmillan.

Waddock, S. (2008). Building a new institutional infrastructure for corporate responsibility. *Academy of Management Perspectives, 22*(3), 87–108.

Waginger, E. (2010). Is there a need to teach knowledge on commodity science and technology at economic universities in a globalized world? *Forum ware, 11,* 219–223.

Wolfgang, H. (2004, August). Austrian education in commodity science—Link between economic ecological and social requests. In *14-th IGWT Symposium "Focusing New Century—Commodity—Trade—Environment"*, Beijing, China.

Elka Vasileva is Associate Professor in the Department of Economy of Natural Resources, University of National and World Economy, Sofia, Bulgaria, where she lectures on standardization, quality and environmental management. Her research is focused on sustainable consumption and production, and consumers policy.

Strategies on Education About Standardization in Romania

Silvia Puiu

1 Good Practices in Education About Standardization

The importance of standards at an international level is widely recognized, these offering a common language between partners from different countries and cultures. Progress related to standards and standardization has been made but there is still room for improvement and education plays an important role here. Education about standardization is the mission of the International Cooperation on Education about Standardization (ICES) founded in 2006, this being the only international entity focused exclusively on this subject. According to the International Federation of Standard Users (2014, p. 4), education is the solution for both the lack of awareness regarding the benefits brought by standardization and for the lack of ability to transform awareness into specific actions.

In Europe, there are three official European Standardization Organizations (ESOs): the European Committee for Standardization (CEN), the European Committee for Electrotechnical Standardization (CENELEC) and the European Telecommunications Standards Institute (ETSI). According to CEN and CENELEC, "education about standardization (EaS) is teaching pupils, students, CEOs, managers, employees in business, and life-long learners the subject of standards and standardization". We can notice that the need for education and training targets not only students but a broad category of people, from pupils to CEOs, highlighting the importance played by standards in our life. The European Commission (2011) asked "ESOs, member states and other standardization bodies to improve awareness and education about standardization".

Hesser and de Vries (2011, p. 10) appreciate that "South Korea is far ahead in implementing standardization education in academic curricula. It does more than Europe as a whole". The University Education Program on Standardization in South

---

S. Puiu (✉)
Faculty of Economics and Business Administration, University of Craiova, Craiova, Romania
e-mail: silviapuiu@yahoo.com

© Springer Nature Switzerland AG 2020
S. O. Idowu et al. (eds.), *Sustainable Development*, CSR, Sustainability,
Ethics & Governance, https://doi.org/10.1007/978-3-030-28715-3_7

Korea started in 2004 and the program for the Elementary School in 2008 (ISO, 2010), thus making this country a model for European ones that need efficient strategies and national policies in the field of education about standardization and standards. ISO (2014, p. 3) highlight the importance played by educational institutions and especially by universities in teaching students about standards and standardization and preparing them as future professionals. The higher education institutions are an intermediate link that can contribute to increasing knowledge on the role of standards but also to improving the existing standards or developing new ones.

According to ISO (2016), one of the outcomes of the Action Plan for developing countries between 2016 and 2020 is strengthen the strategic capabilities of the national standards bodies through different measures, such as: secure the financial stability of these organizations, engage all stakeholders in standardization and promote education about standardization to increase awareness to the role played by standards in our life. The Action Plan specifies that this could be done by improving the relationship with educational institutions and including programs related to standardization in the schools' curricula. The model of South Korea could be used as a starting point for developing programs in all levels of education (elementary school, high and middle school, university and other forms of training beyond formal education).

There is a need for government, standardization organizations and educational entities to work together in order to put the basis for better national policies that stimulate education about standardization and contribute to the creation of future standards experts. According to Hesser and de Vries (2011), Asian countries like South Korea, China and Japan have national programs focused on education about standardization, all these requiring important investments at a national level (funding the courses, teaching the teachers, launching books, workshops and conferences).

The idea of studying standardization is not new. Verman (1973) considered that standardization should be included in the curricula of universities as a standalone discipline. With all the recognized importance of standards, after 40 years, we cannot say that the situation improved much in European universities, at least not to the extent we meet in South Korea and other Asian countries. Spivak and Kelly (2003, p. 24) considered that the "future can be bright" if specific actions are initiated in order to enhance "standards education worldwide", as a common responsibility of all the standards experts, professionals, businessmen, professors and government. A few years before, Hesser and Czaya (1999, p. 9) were already understanding that "the future in teaching has begun" and "the teaching of standardization" had "to adopt a future-oriented approach".

The gap between the Asian model of standards education and the European one is highlighted by de Vries and Egyedi (2007, p. 25). In 2006, there were 46 universities that had standards courses in their curricula only in South Korea, meanwhile in all European countries there were between 10 and 30 universities. In the USA, the situation was even worse, only three universities offering such courses. The solutions seen by the authors are the development of a regional policy and a strong cooperation between government, national standards bodies (NSBs) and educational institutions.

Purcell and Kelly (2003, p. 33) described a situation in the USA where between 1999 and 2001, there was offered a standards course, *Strategic Standardization*, to graduate engineering and law students. The course was intensively promoted and supported by different organizations. Even so, the results were disappointing and only 18 students decided to attend the course in three years. One of the conclusions the authors reached was that students are not attracted by these type of courses if they do not have practical experience.

De Vries (2003) states that standardization should be taught to students in a way they understand and presents learning by examples as a model for standards education. Presenting real examples to students helps them better understand the context in which they will need standards in their professional life and makes them more prepared for the challenges brought by a competitive environment and also one in a continuous change. De Vries (2014, p. 263) highlights that "despite its recent growth", standards education "is an exception rather than a rule". The author also mentions some of the obstacles for such a situation: the subject of standards may not seem very appealing for students and professors may not be willing to teach such a subject because they do not understand the role played by standardization.

Hesser and Siedersleben (2007) consider that the role played by standardization is a consequence of globalization. The international competitive environment makes it necessary for the professionals to have knowledge on standards too. This is how the need for teaching standards appeared. A similar idea is expressed by Lee (2007, p. 17) who argues that the role of standards and the need to know them are increasing as a result of an "intensified global competition and rapid technological innovation". De Vries (2015) appreciates that standards education is also needed for professionals working in NSBs or other international organizations related to standardization and highlights the importance of building a community of these experts in order for them to be able to share knowledge and experience regarding different issues brought by standards.

2 Research Objectives

The objectives of this research are: establishing the role played by standards and standardization, highlighting the need for education about standardization; identifying the barriers to a higher level of knowledge regarding standards; identify the measures, strategies and policies that could be implemented in order to develop education about standards; identify the actors involved and the specific actions that could be done in universities in order to increase the attractiveness of standards courses for students; analyze education about standardization in Romania, with a focus on the need for this type of education in universities and on the knowledge regarding these issues.

The research questions (RQ) we started from are the following:

RQ1: *Which is the level of knowledge about standardization in Romania?*

RQ2: *Which are the strategies that could be implemented in order to increase the level of knowledge, awareness and education about standardization?*

3 Research Methodology

The methodology used for this research is based on: reviewing the professional literature on education about standardization; gather data from ISO, NSBs or ESOs and make a comparative analysis for the member states in European Union regarding education about standardization; conducting a quantitative research on students, professors and other specialists in Romania in order to establish the knowledge regarding standards and the need for educating people on these issues.

The quantitative research is based on a questionnaire that was sent to 600 students or graduates, professors, experts on standardization or other professionals during October and November 2017. We received answers from 194 respondents, having a response rate of 32.33%, which is considered higher than the average (Surveygizmo, 2015), taking into account that the survey was distributed online (by e-mail and social platforms). Even if the target of the survey was represented by people in the higher education system, we also received some opinions like: "a course on standardization would be useless, we need practice not another theoretical course" because we included also some open questions in the questionnaire besides the ones with pre-defined answers.

The limits of an online survey are inherent but taking into account the lack of data regarding EaS in Romania and the slow progress in this direction, this type of research was the one that offered us the possibility to understand the barriers to having a more developed EaS and highlight a few solutions that could be implemented in order to raise the awareness on the importance of this domain.

4 Education About Standardization in Europe

The three European Standardization Organizations created a Joint Working Group on Education about Standardization (JWG-EaS) and developed a Masterplan for Education about Standardization in accordance with the common policy established in 2010 (JWG-EaS, 2011). The main objectives of this working group were to increase the number of people having knowledge on standards and their role, increase the awareness on the need to incorporate standards in the universities' curricula and raise the level of competence for those involved in the process of developing and improving standards.

The Masterplan was intended to define a framework for action, helping national initiatives to introduce standards courses in schools. According to CEN and CEN-ELEC (2011, p. 4), "education and awareness of standards in Europe have not kept pace". The Masterplan also highlights the gap between Asia and Europe and the risk of losing competitive abilities if European countries do not take measures to increase awareness on these issues and implement national and regional policies to support standards education. There were established three main directions for EaS: primary and secondary education targeting school children, tertiary education targeting college and university students and life-long learning for managers and employees. In 2011, the document mentioned that there are some initiatives in European countries but these are "fragmented and the impact is limited" (CEN & CENELEC, 2011, p. 5).

In 2012, the three ESOs organized the first event on EaS within the JWG-EaS—"Strategic importance of education about standardization: a dialogue with academia and industry" (CEN & CENELEC, 2012). The event united representatives from academia, business and experts from the CEN and CENELEC community, being a success. In the following year, CEN, CENELEC and ETSI hosted the conference of the International Cooperation for Education about Standardization (ICES) and the European Academy for Standardization (EURAS), trying to understand the industry requirements for standards education. With the support of CEN and CENELEC, the British Standards Institution, Danish Standards, the National Standards Authority of Ireland, the Finnish Standards Association and the University of Zagreb created a textbook on standards and standardization for higher education in order to increase the awareness on these issues (Bøgh, 2015). The textbook includes case studies, examples, Power Point slides, open and multiple choice questions.

Even if the activities developed by the JWG-Eas were diverse and contributed to the increase of knowledge on standards, this working group was revised at the end of 2016 and as a consequence it was disbanded without having finished its activities, being considered that the NSBs are "very familiar and acquainted of the national standardization and education systems" (CEN & CENELEC, 2017).

According to a survey conducted by Egyedi (2009), there were 84 European entities active in standards education at that time, with no activities in primary education, nothing systematic at the secondary level (with the exception of British Standards Institution that addresses to all ages and educational levels) and almost 70% of the entities being higher education institutions. The study also highlights the Austrian Standards Institute which created a CD for secondary level teachers in order to increase their knowledge related to standards. Egyedi concludes that the courses entirely focused on standardization are considered "extravaganza", being an exception at least for the undergraduate level. The countries mentioned in this report as having a national strategy for standards education are United Kingdom, France and Germany. The Netherlands also plays an important role in this field but doesn't have a specific strategy. The most advanced country regarding standards education is United Kingdom where British Standards Institution (BSI) offers resources for each educational level, both for teachers and students, in accordance with the level of knowledge and the age of each beneficiary.

Hesser and de Vries (2011) also presents the situation of standards education in Europe, highlighting that Europe plays an important role in the standardization research community. The authors mention the efforts of EURAS to stimulate education about standardization. At a national level, Hesser and de Vries present the name of some universities in France and Germany that included standards courses in their curricula and the situation in the Netherlands where the Netherlands Standardization Institute supports standards education at Erasmus University and a network of researchers in standardization comprised of 9 of the 12 Dutch universities. In Hamburg at Helmut Schmidt University, there was an optional standards course since 1984 and Professor Hesser developed an e-learning platform for this subject (www. pro-norm.de). In the Netherlands, Jan Smits at University of Technology Eindhoven asked his students to write a blog on standardization and Tineke Egyedi developed at Delft University of Technology, in collaboration with United Knowledge, a simulation exercise "Setting Standards" for students and policy makers.

Hesser and de Vries (2011, p. 17) state that "there is a big discrepancy between policy and practice" and that "current practice shows no more than fragmented standardization education activities in the EU and hardly any programmes at the academic level". Between the research of Egyedi (2009) and that of Hesser and de Vries (2011), there are two years but the results and the conclusions are similar. We can say that there is not much progress regarding education about standardization in Europe. The exceptions represented by UK, Germany, France and the Netherlands are recognized by both papers.

Whitney, Keith, Selvaraj, Maguire, and Nicolle (2014) presents the facilities offered by BSI for students, teachers and other persons that are interested in standards with the aim of raising the awareness on the role played by standardization in our lives: British Standards Online (BSOL) is an online standards database that comprises more than 60,000 national, European and international standards; BSI Education section created for students in higher education and providing them useful information; BSI shop for buying books and publications about standards; BSI Speakers Network that offers talks about standards; BSI Standards Development website which encourages people to check the standards in progress and offer suggestions for improvement. The authors conducted three surveys (on students, academics and specialists from the private sector) related to the standards education and the need for such courses. Some of the conclusions the researchers reached were: the academics highlighted the importance of correlating the theory on standards with practical examples, case studies and assignment in order to increase the attractiveness of standards courses and make them more useful for students; most students were engaged with one or more standards, two thirds of them had full access to standards using BSOL or other online resources; the students wanted to be able to apply standards in a specific context in real life; the employers recognized the importance of standards for their business (almost 75% of them) and 50% of them were also involved in the development of national or international standards; most of the employers said they required standards knowledge from students but they also offered them the possibility to learn about this topic at work.

The results of the three surveys offer us an image of the standards education in the UK, which is the most advanced so far in Europe in terms of having a strategy on standards education and also an online database for standards. The UK and BSI are examples of good practice for the countries in Europe. Of course there are also some issues in this system, some of the employers in the above survey considering that physical copies of standards are more reliable than online sources and that academic should prepare students better in order not to increase the cost for the employer.

In Denmark, the national standardization body is also very active, Danish Standards Foundation (DS) having a great history behind, being founded in 1926 and also known for the two eco-labels—the European Flower and the Nordic Swan. DS published a textbook for higher education in collaboration with the British Standards Institute, the National Standards Authority of Ireland, the Finnish Standards Association and the University of Zagreb within the JWG-EaS. There are four types of activities developed at DS: standardization, courses and conferences, sale of standards and textbooks and eco-labelling. Hesser (2014, p. 3) analyses the situation of EaS in European countries, highlighting that "development of the educational system in Denmark may be regarded as unusual" because there was a great increase of the number of courses focused on standards and taught to students—1760 students completed 177 courses in 2012 comparing with 1290 students and 86 courses in 2010. In 2014, United Knowledge and the Delft Institute of Research on Standardization created three simulation games for the DS: Good Teaching! (about the meaning and role of standards), The Sky's the Limit (standards and innovation) and Allprod Incorporated (Egyedi, 2017).

According to ISO (2014), NSBs could contribute to the progress of standards education no matter what barriers they encounter if they have a good collaboration with the universities and they set objective goals on this subject. The report states that there are a few obstacles to an efficient partnership between NSBs and higher education institutions: standardization education is not considered a priority and there are few specialists willing to dedicate their time and efforts to this type of projects or they do not have the needed knowledge; the connection between education and standardization is not always seen as important; lack of financial resources for supporting these initiatives; a lack of interest for standardization in universities; lack of support from the government and other public authorities; lack of teaching materials and experience; limited availability of specialists from NSBs, industry or universities to contribute to standards educational programs as trainers. There is a high need for offering students examples and practical assignments in order for them to be able to better understand the contents of the standards they learn about.

The same report (ISO, 2014, p. 87) states that there are six success factors for improving the collaboration between NSBs and universities in order to introduce standards courses in the academic curricula and raise the awareness on this topic: considering standards education a priority; finding value in standardization education and understand the role that it plays in the society; standards education in universities should be focused on the needs of the market; a strong link with the ministries of education and industry in order to have their support; engagement of stakeholders to have their support (for example companies who are interested in having competent employees in terms of standardization knowledge); exploit the standardization network at a local, regional, national or international level.

5 Education About Standardization in Romania

The literature on the subject of EaS in Romania is not so abundant. We can find only a few articles and courses regarding this topic on the website of ASRO, the Romanian standards body, and a couple of other papers written in Romanian (Bejan, 2012; Martis, 2014). Bejan (2012) makes a short synthesis of the history of standards at the national and international level, presenting the benefits of standards in our lives. He presents his opinion as an engineering professor and highlights the need for lifelong learning.

The author offers a few solutions for increasing the awareness on the activities initiated by ASRO: strengthening the relation with higher education; developing new studies and promoting the old ones in order to raise the awareness of the role played by standards; development of educational programs on standardization in different areas like engineering, management; creating a communication network for all stakeholders and encourage the engagement of academics in the process of developing standards. The benefits for introducing standards courses would be a qualified and competitive workforce.

Martis (2014) presents the benefits of EaS on business (having knowledge on standardization leads to a competitive advantage), authorities (standards contribute to a common technical language and increase the quality of products and services), market (competitiveness, a more qualified workforce) and students (they will be better prepared for the labor market). The solutions provided by the author to increase the awareness of the role of standards are: establishing the educational needs; creation of a working group with representatives of all stakeholders (industry, NSB, government, higher education); elaborating an action plan; establishing the individuals who will be in charge with the plan; identifying the needed resources; developing textbooks and other educational resources; developing a program for training the teachers; promotional activities for these initiatives.

On the website of ASRO, we can find a presentation of the concept "education about standardization", including references to the policy of CEN and CENELEC on EaS (translated into Romanian), to their masterplan (only in English) and implementation plan (only in English).

An online search using terms like "standards courses in Romania", "standards education in Romania", "education about standardization in Romania" both in Romanian and in English shows us the lack of research on this topic and that very few universities have in their curricula an independent course on standards and standardization (Faculty of Chemistry, University of Bucharest includes a course on standardization and legislation in chemistry; Faculty of Electronics, Telecommunications and Information Technology, University Politehnica of Bucharest includes a course about standardization and legislation in quality and safety). Most economic and engineering faculties have elements of standardization in a broader course, but not one that is entirely focused on standardization.

ASRO offers ten e-learning courses on standardization in general targeting all specialists in the industry that need knowledge on this topic in order to increase their competitiveness. There are also two courses on standards development: one is introductive and the other one is more advanced, being focused on the working procedures in the Technical Committees, in ISO, CEN, IEC and CENELEC. For the new members in the Technical Committees there are offered free courses on a monthly basis in order to raise their skills in standardization.

In order to have a clearer picture on education about standardization in Romania, we conducted quantitative research based on an online survey, which was distributed via e-mail and social platforms to 600 students, professors and other specialists in the industry, public entities or experts from ASRO. We focused mainly on students and professors in the economic field in Romania and sent the survey to ASRO in order to help us with dissemination. Because we did not benefit from the support of ASRO (only one member responded), we acknowledge the limits of this research and that the sample might not be representative for drawing more general conclusions. But the results are useful for specialists in this field and could help other researchers in their efforts to talk about EaS in Romania. The questionnaire was available during October and November 2017 and the response rate was 32.33% (194 respondents).

The respondents consisted of 60.3% students, 21.1% teachers, 7.2% specialists in the private sector, 2.1% specialists in the public sector and 9.3% other categories (graduates, specialists in ASRO, managers). We contacted ASRO by e-mail and social platforms in order to support our research disseminating the survey and also for answering the questionnaire. They responded in a supportive manner but only one person from the NSB responded to the survey.

Asked if they taught or were taught a course including aspects related to standardization in higher education, most of the respondents (62.4%) said no. Those who responded yes (37.6%) mentioned the domains in which they had a course with elements of standardization: economic field (86%), technical field (8%) and other domains like IT, ecology, and food (6%). Most of the respondents (67.1%) mentioned higher education entities as the main provider for these courses, but there were also 17.8% who mentioned a private entity (mainly companies who send their employees to training), 12.3% other public entities and 2.8% NGOs.

When the respondents were asked about the need to introduce a standards course in the academic curricula, 93.1% agreed that such a course should be taught in faculties. The high percentage show us that students and professors understand the role of standards and this is a good background for further strategies on EaS. In many situations, standards courses are optional and do not seem attractive for students so they do not choose them, as in the study of Purcell and Kelly (2003) in which a course on standards was offered for students in law and engineering and despite its promotion, between 1999 and 2001, only 18 students chose to attend the course.

In this context, we asked respondents to mention the measures that can be implemented in order to increase the attractiveness of these courses. The data from Table 1 show us that the most important solutions were: studying examples of good practices at an international level (57.7%), understanding the way standards contribute to an increase in the quality of life (44.8%), understanding principles of standardization (38.1%), understanding the link between standardization and competitiveness (37.1%), and learning about the advantages brought by standardization (35.6%).

Table 1 Solutions to increase attractiveness of standards courses

Solutions	Percentage (%)
Studying examples of good practices at an international level	57.7
Understanding the way standards contribute to an increase in the quality of life	44.8
Understanding principles of standardization	38.1
Understanding the link between standardization and competitiveness	37.1
Learning about the advantages brought by standardization	35.6
Analyzing the content of various international standards	30.9
Understanding the way in which the standardization system can be improved	22.7
Studying about national and international standardization organizations	20.6
Studying the evolution of the standardization process	12.9
Other (applicability, learning how to implement the good practices)	1

We can notice that the respondents are attracted to standards courses when they have a practical component, helping students to understand how standards contribute to a better life and a higher competitiveness. Studying about the evolution of the standardization process is important but not so interesting for students. Learning about the good practices at an international level seems to be the most important aspect that should be taken into account when developing and implementing a national strategy on EaS.

Besides formal standards courses in the academic curricula, the respondents mentioned that there are also other solutions that could be efficient for learning about standards and standardization, as we see in Table 2.

Most of the respondents considered workshops and conferences as the best way to learn about standards by communicating and sharing ideas with other specialists (58.2%), closely followed by internships and trainings (53.6%). Correlating these results with the previous question, we can understand why these types of activities are more practical and efficient as other courses, probably seen as being more theoretical. Thus, things can be improved and EaS in higher education can lead to more attractive and interesting courses with an important practical component.

Table 2 Other solutions for learning about standards

Solutions	Percentage (%)
Workshops/conferences for meeting other specialists	58.2
Internships/trainings at national or international standardization organizations	53.6
Online courses	27.8
Intensive courses focused on a specific area of standardization	24.7
Individual study on standards	13.4
Postgraduate courses	11.9

Strategies on Education About Standardization in Romania

When they were asked if standards courses should be mandatory for students, most of the respondents (56.7%) agreed, followed by those who had a neutral opinion (31.4%). Only 11.9% of the respondents disagreed with the idea of the courses being integrated in the academic curricula. So, there is an interest in these courses and the problem is in fact with the way the courses are presented to students which makes all the difference between efficient and non-efficient. In order to see the preference of the respondents for more general or specific standards courses, we used a Likert scale. The results in Table 3 show us that people would prefer an equilibrium between general knowledge on standardization and specific knowledge focused on a particular area of interest.

If we look at the results in comparison, we notice that the respondents would prefer an equilibrium between general and specific topics in standards courses (77.9% total or partial agreement) but if they have to choose between general and specific, they would choose having more general knowledge (52.6% total or partial agreement), probably having in mind that the labour market is very dynamic and it would be useful to have a broader view on standardization.

Asked if EaS should be translated into national policies and strategies regarding education, most of the respondents agreed with this statement (70.1%), 24.2% were neutral and 5.7% disagreed. Romania does not have a national strategy on EaS and this should be seen as a priority by officials. If we look at examples of good practices in Europe, we can learn from UK (Whitney et al., 2014) adopting some of the measures they successfully implemented in all levels of education, but mostly in higher education. Regarding the need to train the employees by offering them a course on standardization, most of the respondents mentioned there is a need for these trainings (84.5%), followed by 12.4% who did not considered them important and 3.1% who agreed the role of these courses for the employees but depending on the job and only if the job requires standards knowledge. The respondents were also asked about the best length of a standardization course and at which level of education should it be included. We see in Table 4 that opinions were quite heterogeneous.

Even if most of the respondents considered stand-alone standards courses as a better option (51%) than a few chapters on standards in other courses (49%), the difference between these opposite opinions is quite small and we cannot draw a relevant conclusion. Most of the respondents also considered that these courses should

Table 3 General or specific standards courses

Agreement/disagreement	Standards courses should be more general (%)	Standards courses should create an equilibrium between general and specific knowledge (%)
Totally agree	19.1	45.9
Partially agree	33.5	32
Neutral	35.6	19.6
Partially disagree	8.8	1.5
Totally disagree	3.1	1

Table 4 Length and level of education for standards courses

	Percentage (%)
Length of standardization courses	
A stand-alone course during an entire semester	51
A few chapters in other courses	49
Level of education	
Bachelor's degree	34.5
Master's degree	27.3
Both	38.1

be included in the curricula for both the bachelor's and master's degree (38.1%), followed by those who chose only bachelor's degree (34.5%) and those who chose only the master's degree (27.3%). Table 5 reveals a worrying situation regarding the number of people who know about the Romanian NSB—ASRO. They were asked if they know about the existence of such an organization in Romania.

Most of the respondents (53.1%) said they do not know if such an organization exists in Romania and we can assume that the percentage of those who said that there is no such entity (9.8%) could be summed up to the previous one resulting in 62.9% of the respondents who are not aware of the existence of ASRO.

Asked if the standards courses should have a similar structure with the courses in other European countries, most of the respondents (76.3%) agreed (totally or partially) with this statement. The average score for the Likert scale is 1.23 (agreement). Only 1% of the respondents disagreed and the rest of them (22.7%) were neutral.

Taking into account the attractiveness of massive open online courses, the respondents were questioned about their willingness to enroll in a free standardization course that would be available for all European countries, but only in English, not in Romanian. Most of the respondents said they would enroll in such a course (74.7%), followed by those who said they would if they improved their English (14.4%), so an impressive percentage of respondents (89.1%) are willing to learn new things about standards and standardization. Only 10.9% of the respondents are not interested in these type of courses.

The results of this research help us understand the level of knowledge that students, professors and other specialists in the industry in Romania have regarding standards and standardization. All these data correlated with those provided from the website of ASRO and those gathered from the literature on the situation of EaS in Europe

Table 5 Knowledge on the existence of a NSB

Does Romania have a national standards body?	Percentage (%)
Yes	37.1
No	9.8
I do not know	53.1

contribute to the development of a strategy that could be implemented in order to raise the level of knowledge and increase the country's competitive advantage on the market.

First of all, there now is a strategy on standardization in Romania which was adopted in 2014 (ASRO, 2014). The objectives of the strategy include also the need to promote EaS and the activities of ASRO at a national level to strengthen its image. Still, the strategy covers the period between 2014 and 2020, but not many initiatives are visible yet in the field of EaS and many people from the higher education are not even aware of the fact that this organization exists. ASRO is not the only responsible entity, but also the Romanian Agency for Quality Assurance in Higher Education and the government which have to cooperate in order to include EaS in the strategy on education, with a focus on higher education.

Other measures should include partnerships with universities to stimulate the interest for standards and create materials for teaching standards in an attractive manner and thus assuring a better visibility of ASRO. The Romanian Agency for Quality Assurance in Higher Education may make it mandatory for faculties to include standards courses in their curricula.

6 Conclusion

EaS is a responsibility of many actors: the government, the Parliament, the NSB, higher education institutions, professors, employers. All stakeholders should cooperate in order to raise the level of knowledge on standards. Strategies on standardization and strategies on EaS are a challenge for most countries. Examples of good practices could be used to create an adequate background that makes it easier to include EaS in all levels of education.

To have knowledge on standardization is important mainly because we live in a global world, where in order to be competitive, someone needs to speak a common language and standardization could be this language. Higher education institutions have a great role in educating future professionals, but other NGOs and the industry itself could have their role too in increasing the awareness on the importance of standards.

If we look at the model of successful countries in terms of having a high level of EaS, we can say that it is better to address these aspects in all levels of education. The need for having standards courses in the academic curricula is real and the level of knowledge on these issues may be seen as an obstacle but with adequate measures, policies and strategies, things can change. These courses should be more attractive for students incorporating examples of good practices and having a practical component.

The strategies that could be used in Romania in order to ensure a proper EaS are: creating a network for helping stakeholders to communicate on these issues; organizing workshops and conferences on standards and promote them within the community; introduce standards courses in all levels of education; a partnership between ASRO and schools to create teaching materials, presentations and an exchange of information; developing a database on standards that could be accessed by students, professors and other specialists; national campaigns to promote the role of standards in our lives and raise the awareness of the population on these problems; an active presence of the country within ISO and all the European Standardization Organizations.

EaS combines the education and standardization domains, reuniting strategies specific to both of them. Strategies adopted by countries like South Korea or United Kingdom should be thoroughly studied in order to understand how these could be implemented in other countries too without ignoring the national context. Progress in EaS is a prerequisite for becoming competitive at an international level for both individuals and countries.

References

ASRO. *E-learning courses on standardization*. Retrieved from http://www.asro.ro/?page_id=940.

ASRO. *Standardization courses*. Retrieved from http://www.asro.ro/?page_id=967.

ASRO. *Standardization courses for the members in the technical committees*. http://www.asro.ro/?p=3377.

ASRO. (2014). *Strategia privind activitatea de standardizare nationala in Romania*. Retrieved from https://issuu.com/asociatiadestandardizaredinromania/docs/strategia-asro-2014-2020_aprobata-d.

ASRO. (2015). *Educatie despre standardizare*. Retrieved from http://www.asro.ro/?page_id=1527.

Bejan, M. (2012). *Relatia standardizare—Mediul universitar*. Univers Ingineresc no. 2/2012. Retrieved from http://www.agir.ro/univers-ingineresc/numar-2-2012/relatia-standardizare—mediul-universitar_3575.html.

Bøgh, S. A. (Ed.). (2015). *A world built on standards—A textbook for higher education*. Danish Standards Foundation. Retrieved from http://uni.ds.dk/~/media/DS/Files/Downloads/Uni/A-World-Built-on-Standards.pdf.

CEN & CENELEC. *Education and training on standardization*. Retrieved from https://www.cencenelec.eu/standards/Education/Pages/default.aspx.

CEN & CENELEC. (2011). *Masterplan for education about standardization*. Retrieved from https://www.cencenelec.eu/standards/Education/JointWorkingGroup/Documents/Masterplan%20on%20Education%20about%20Standardization.pdf.

CEN & CENELEC. (2012). *Strategic importance of education about standardization: a dialogue with academia and industry*. Conference report. Retrieved from https://www.cencenelec.eu/standards/Education/JointWorkingGroup/Documents/ReportConference2012.pdf.

CEN & CENELEC. (2017). *Joint working group on education about standardization*. Retrieved from https://www.cencenelec.eu/standards/Education/JointWorkingGroup/Pages/default.aspx.

De Vries, H. J. (2003). Learning by example—A possible curriculum model for standardization education. ISO Bulletin July 2003. Retrieved from https://www.iso.org/files/live/sites/isoorg/files/archive/pdf/en/article_bull_july_03_-_learning_by_example.pdf.

De Vries, H. J. (2014). How to implement standardization education in a country. In K. Jacobs (Ed.), *Modern trends surrounding information technology standards and standardization within organizations*. IGI Global.

De Vries, H. J. (2015). *Building a community of standards professionals*. Paper presented to the 2014 conference of the ICES and published in cooperation with ICES by ISO in January/February 2015. Retrieved from http://www.iso.org/sites/edumaterials/ses-community-of-professionals-2015.pdf.

Strategies on Education About Standardization in Romania

De Vries, H. J., & Egyedi, T. M. (2007). *Lessons from Asia: Bridging the gap between theory and practice*. ISO Focus November 2007. Retrieved from https://www.iso.org/files/live/sites/isoorg/files/archive/pdf/en/iso_focus_nov_2007_-_bridging_the_gap_bertween_theory_practice.pdf.

Egyedi, T. M. (2009). *The state of standards education in Europe*. ICES Workshop, Tokyo, 23–24 March 2009. Retrieved from http://www.standards-education.org/uploads/ices2009/European_Standards_Educ_ICES2009_EgyediFinal.pdf.

Egyedi, T. M. (2017). Teaching standardization in engineering, science, and technology studies. Retrieved from https://www.standardsuniversity.org/e-magazine/december-2017/teaching-standardization-engineering-science-technology-studies/.

European Commission. (2011). *A strategic vision for European standards: Moving forward to enhance and accelerate the sustainable growth of the European economy by 2020*. Communication COM/2011/0311 final. Retrieved from http://eur-lex.europa.eu/legal-content/EN/TXT/?uri=celex%3A52011DC0311.

Faculty of Chemistry, University of Bucharest. *Standardization course in chemistry*. Retrieved from http://www.unibuc.ro/prof/buleandra_m/docs/cvs/2016ianmihaela_buleandra.pdf.

Faculty of Electronics, Telecommunications and Information Technology, University Politehnica of Bucharest. *Standardization course in the master's curricula*. Retrieved from http://www.euroqual.pub.ro/cursuri/standardizare-si-legislatie-in-calitate-si-siguranta-in-functionare/.

Hesser, W. (2014). Memorandum on standardization in higher education in Europe. Haburg, Germany. Retrieved from https://www.iso.org/sites/edumaterials/hesser-memorandum.pdf.

Hesser, W., & Czaya, A. (1999). *Standardization as a subject of study in higher education: A vision*. ISO Bulletin June 1999. Retrieved from https://www.iso.org/files/live/sites/isoorg/files/archive/pdf/en/article_bull_june_99_-_stdzation_study_higher_education.pdf.

Hesser, W. & De Vries, H. J. (2011). *Academic Standardization in Europe*. EURAS. Retrieved from http://www.iec.ch/about/globalreach/academia/pdf/academia_governments/EURAS_White_paper_2011-08-13.pdf.

Hesser, W., & Siedersleben, W. (2007). *Standardization goes East*. ISO Focus November 2007. Retrieved from https://www.iso.org/files/live/sites/isoorg/files/archive/pdf/en/iso_focus_nov_2007_-_article_the_european_asian_academic_network.pdf.

International Federation of Standard Users. (2014). *Education and training about standardization*. Retrieved from http://www.ifan.org/IFAN-GUIDE%204-Education-2014-09.pdf.

ISO. (2010). Education securing Korea's future. Retrieved from http://www.iso.org/sites/WSCAW2010/materials/presentations/07-07-wednesday/02-Afternoon/3-WSC-KATS-final.pdf.

ISO. (2014). *Teaching standards: Good practices for collaboration between national standards bodies and universities*. Retrieved from https://www.iso.org/files/live/sites/isoorg/files/publications/en/teaching_standards_pub100354.pdf.

ISO. (2016). *ISO action plan for developing countries 2016–2020*. Retrieved from https://www.iso.org/files/live/sites/isoorg/files/publications/en/action_plan_devel_2016-2020.pdf.

JWG-EaS. (2011, February). Policy on education about standardization. Retrieved from ftp://ftp.cencenelec.eu/EN/EuropeanStandardization/Education/Policy%20on%20EaS.pdf.

Lee, G. H. (2007). *Universities in the Republic of Korea: Training the next generations of professionals*. ISO Focus November 2007. Retrieved from https://www.iso.org/files/live/sites/isoorg/files/archive/pdf/en/iso_focus_nov_2007_-_article_training_next_generation.pdf.

Martis, M. (2014). *Strategic importance of education in the field of standardization*. In The XIV International Conference "Profesorul Dorin Pavel—fondatorul hidroenergeticii romanesti", Sebes. Retrieved from http://stiintasiinginerie.ro/wp-content/uploads/2014/07/25-14.pdf.

Purcell, D. E., & Kelly, W. E. (2003). *Adding value to a standards education: Lessons learned from a strategic standardization course*. ISO Bulletin July 2003. Retrieved from https://www.iso.org/files/live/sites/isoorg/files/archive/pdf/en/article_bull_july_03_-_lessons_learned.pdf.

Spivak, S. M., & Kelly, W. E. (2003). *Introduce strategic standardization concepts during higher educational studies... and reap the benefits*. ISO Bulletin, July 2003.

Retrieved from https://www.iso.org/files/live/sites/isoorg/files/archive/pdf/en/article_bull_july_ _03_-_strat_stdzation_concepts.pdf.

Surveygizmo. (2015). *Survey response rate*. Retrieved from www.surveygizmo.com/survey-blog/ survey-response-rates/.

Verman, L. C. (1973). *Standardization: A new discipline*. Archon Books.

Whitney, G., Keith, S., Selvaraj, N., Maguire, M., & Nicolle, C. (2014, May). *Employable knowledge: Benchmarking education about standardization in the UK*. British Standards Institution. Retrieved from https://www.bsigroup.com/LocalFiles/en-GB/standards/BSI-Benchmarking-Education-about-Standardization-UK-EN.pdf.

Silvia Puiu is a Ph.D. Lecturer at the Department of Management, Marketing and Business Administration within the Faculty of Economics and Business Administration, University of Craiova, Romania. She has a master's degree in International Business Administration and a Ph.D. in Management at University of Craiova. She published her Ph.D. thesis on Strategic Management of the Retail Sector in Romania in 2012. In 2014, she conducted a research on *Ethics Management in Higher Education System of Romania* and in 2015, she graduated her postdoctoral studies on *Ethics Management in the Public Sector of Romania*. Silvia teaches Management, Business Ethics, Public Marketing and Creative Writing. She published more than 40 articles in national and international journals, two books and a chapter—*Corporate Social Responsibility in the Romanian Public Sector*—in the book *Corporate Social Responsibility in Times of Crisis* (Springer, 2017). Her research covers topics from strategic management, ethics management, public marketing, corporate social responsibility and management. She is also a reviewer for two journals and a member of Eurasia Business and Economics Society.

Stimulating Education About Standardisation

Necessary Competences of Employees in the Field of Standardization

Knut Blind and Sandra Drechsler

1 Introduction

According to ISO *"standardization is a rich and complex domain, which requires sound technical and business knowledge, combined with soft skills (interpersonal communication, behaviour in a consensus-oriented—often international and multicultural—environment, negotiation and lobbying in that context), knowledge about public policies in selected areas and, in some cases, legal competence"* (ISO, 2014, pp. 18–19). Hence participation in standardization processes is a complex task, exceeding purely technical expertise. It requires to know and to understand legal, political, organizational and personal characteristics and relationships. A lack of awareness in these conditions can lead to mistakes with far-reaching consequences for the respective company (Cooklev & Bartleson, 2008). Some senior managers from companies actively indicate a need to carefully balance pre-competitive work, helping to set frameworks benefiting all market players, with key value adding activities in which their companies seek competitive advantage (ISO, 2014). Consequently, this activity places high requirements on the employees concerned regarding their knowledge, skills and competences. Thus, they should have knowledge in the following fields: standardization organisations, standards and legislation, standards in global markets, standards and intellectual property rights, product development strategies, costs and advantages of standardization as well as risks and advantages of standards (Kurokawa, 2005), existing and future regulatory framework conditions but also about the research, innovation and knowledge management of their company

---

K. Blind (✉)
Faculty of Economics and Management, Fraunhofer Institute for Open Communication Systems FOKUS, Technische Universität Berlin, Berlin, Germany
e-mail: Knut.Blind@TU-Berlin.de; Knut.Blind@fokus.fraunhofer.de

S. Drechsler
Institut Für Produktentwicklung, Karlsruher Institut Für Technologie (KIT), IPEK, Karlsruhe, Germany
e-mail: Sandra.Drechsler@vdma.org

© Springer Nature Switzerland AG 2020
S. O. Idowu et al. (eds.), *Sustainable Development*, CSR, Sustainability, Ethics & Governance, https://doi.org/10.1007/978-3-030-28715-3_8

(Blind & Mangeldorf, 2016). In addition, soft skills are highly relevant, such as communication skills, ability to generate consensus, assertiveness, and abstraction capability, transferability, negotiating skills and the ability to estimate consequences of one's own actions (Drechsler & Albers, 2016). As standards are highly significant for the economy, there is need for special education, which is expected to even grow in the future (Hövel & Schacht, 2013), (The Center for Global Standards Analysis, 2008). According to (Carugi, 2013) students as well as companies would benefit from education in standardization as necessary competences and skills such as communication, negotiation and analyses can best be acquired in an academic environment, like universities. Especially the „*awareness of technical and business trends, [as well as the] competitive intelligence [and the] suppliers' evaluation*" should be mediated (Carugi, 2013, p. 4).

Even the European Commission (EC) has repeatedly pointed out the need to improve the education with regards to standardization (European Commission, 2011), a need confirmed by the "Joint Initiative on Standardization (JIS)" of the public and private partners in the European Standardization System (European Commission, 2016). To explore and promote standardization as an element of formal, academic as well as vocational training a comprehensive assessment of market demand is required. Thus, it has been necessary to identify most relevant areas for standardization-related education as well as job specific requirements in standardization knowledge, skills and competences. This task has been carried out within a European study on market needs for education in standardization (Blind & Drechsler, 2017).

The aim of this paper is to provide an overview of the job profiles related to standardization, and job specific requirements in standardization knowledge, skills and competences.

This contribution is organised as follows. After a review of the literature, we present the analysis of current job profiles related to standardization in Europe. Based on a survey among industry experts we developed a general competence model for standardization. Finally, we conclude with a summary and some recommendations.

2 Literature Review

Two different approaches have been chosen for the literature analysis to ensure that all relevant documents and publications on the subject are considered. On the one hand this was achieved by searching for all academic literature on education in standardization by consulting the databases Web of Science provided by Thomson Reuters, Scopus provided by Elsevier and Google Scholar. On the other hand, all relevant policy documents and non-academic reports, white-papers and press releases (grey literature) have been reviewed, e.g. via internet searches. The following is a summary of the most important results from two approaches literature review, (Blind & Drechsler, 2017).

Companies usually have permanent and rotating personnel active in committees. The rotating personnel are technical experts who are seconded to committees if a standard directly influences their respective field of activity. (Bailetti & Callahan, 1995) Hence, employees, who are participating in committees, can be either (Drechsler & Albers, 2016):

- Employees whose main task is the field of standardization, both internally and externally. Their position can be located within a standardization department or as a staff position, designated as standardization expert in the following. (see Chap. 2.1)
- Employees who mainly work in technical departments for example and who are involved in standardization issues in their specific field of activity as they are the experts for a certain technical matter (Chap. 2.2).

Recently, there has been a trend that experts have withdrawn from committee work (Nagel, 2002). The results of the German standardization panel show that although companies have withdrawn somewhat from formal standardization since 2015, they have instead become more involved in consortia (Blind & Müller, 2017). Persons active in the committees has been mostly engineers because discussions in committees are highly technical (Weiss, 1993), but there is a trend that in fact fewer and fewer developers and designers are represented in the committees and the work has increasingly being taken over by sales engineers and marketing specialists (Nagel, 2002). Since the technical level of a solution combined with the technical knowledge of the person presenting it has a great influence, persons without a technical background are no suitable company representatives (Weiss, 1993). These representatives should be technical experts with a high degree of experience-based knowledge (Drechsler & Albers, 2016, McMillian, 2013, Jakobs, Procter, & Williams 2001). On the strategic level, however, experience either in standardization or in company's strategy is more important (Kurokawa, 2005).

Apart from that, the activity standardization places high demands on the employees concerned regarding their knowledge, skills and competences, such as (Tanaka, 2010)

- Standard development: type of standard (e.g. products/ performance), functions (e.g. compatibility), procedures of standard writing, drafting standards, legal framework, knowledge of foreign standards
- Standard implementation: relation to other standards released by international SDOs or consortia, Knowledge of the glossary (e.g. ISO/ IEC, ASTM), search tools for standards, conformity assessment (e.g. CASCO tool box IAF).

Consequently, speaking about market needs for education-related competences requires having a deeper look in, which job profiles currently exist in standardization, which kind of education is required, and which competences are assigned to these particular job profiles from a business point of view. For such an analysis, career profiles that have a direct reference to standardization must be taken into account, as well as profiles that only have a certain proportion of standardization in addition to the actual activity, (Blind & Drechsler, 2017).

Within companies' employees with different degrees carry out the same jobs and employees with similar degrees different jobs. Consequently, we can either differentiate by position/ department or by educational background. In the following we take the company perspective while the university perspective is not part of the following consideration, (Blind & Drechsler, 2017).

2.1 Job Descriptions in Relation to Standardization

Within this section we give a brief overview about the research regarding job descriptions and their particular demand on knowledge, skills and competences of employees whose main task is the field of standardization. As mentioned in the introduction of this chapter their position can be located within a standardization department or as a staff position, designated as standardization expert in the following.

The state of research delivers five job descriptions to describe the position and activity of these employees, cf. Figure 1. At this place, only job profiles are listed which are directly related to standardization and also contain the reference in the title.

The terms *Standards-Manager*, *Standard-Developer* and *Standards-Engineer* were introduced by (Freericks, 2013) in order to describe job descriptions relating to standardization. Even though any further description of specific tasks and respon-

Standard Manager
Planning and implementing a company's standardisation strategy
Providing information about relevant standards

Standard Engineers
range of activities varies between committee work to pure technology-oriented tasks (standard application)

Standard Developer
Committee work

Standard Generalist
Standard administration and management, Company standards, Internal trainings, consultancy in matters of standards, committee work

Standard Specialist
Consultancy in matters of standards, committee work

Fig. 1 Job descriptions related to standards (Blind & Drechsler, 2017, p. 35)

Necessary Competences of Employees in the Field ... 117

sibilities are missing in her contribution, these terms have partly been discussed by other researchers as shown below. Competences regarding standardization, which have been defined by other authors, are added to the above descriptions if referencing the same activity, (Blind & Drechsler, 2017).

2.1.1 Standard Manager

A standard manager is a technical specialist experienced in planning and implementing a company's standardization strategy. These managers are able to recognize the value of external information, to harness this information for economic purposes and to influence conditions and timing of standardization. The tasks of standard managers within companies are the design and implementation of standard strategies including providing information about standards in order to consider them in product development. Contents of standards can only effectively be influenced by an active participation in standardization processes (Bailetti & Callahan, 1995). Standardization managers who are often either embossed by technical or in legal environments have to be aware of the multi-dimensional motives to standardise (Blind & Mangelsdorf 2016), whereby the pure technical aspects are of relatively minor importance. The high relevance of the motives related to regulation reveals that companies active in standardization are not only developing common specifications, but are also trying to influence the regulatory framework. The common interests inherent in engagement with formal Standards Developing Organizations (SDOs) must be considered with respect to individual interests of participating firms. The multidimensional motives to participate in standardization require that standardization activities are embedded in an organizational setting that allows the cooperation between the R&D departments, innovation management, quality management and units responsible for regulatory affairs (Blind & Mangeldorf, 2016). Consequently, standard managers can also be involved in human resource management, patenting activities and institutional arrangements (Bailetti & Callahan, 1995).

2.1.2 Standard Engineers

For the professional title of a standard engineer no unique description exists so far. Their responsibilities as well as necessary knowledge, skills and competences heavily depend on the profile of the respective company, because the range of activities varies between representations of the company in external committees to pure technology-oriented tasks (Freericks, 2013). (Hesser & de Vries, 2011) take the term up and attribute *standard engineers* the following competences: "*(a) being aware of interpreting standards, (b) standards development, (c) standards implementation, (d) standards compliance/ certification/ inspection/ evaluation, (e) standards diffusion/ training*", combined with skills such as for example "language skills" or "ability to build consensus". These standards engineers are usually technical experts for the development of standards in standardization committees. As no appropriate education exists so far, up to the present these employees are chosen based on professional experience and certain soft skills.

2.1.3 Standard Generalist

Standard generalists are employees, who are responsible for standard management and administration, the purchase of standards, the administration and maintenance of company standards and the provision of internal trainings. Standard generalists advice and support technical departments in matters of standards. They represent the own company in external committees and in some companies this task is carried out by former employees of technical departments, (Drechsler, 2016).

2.1.4 Standard Specialist

Standard specialists are technical experts representing the own company in specific committees, advising technical departments in matters of standards and reviewing customer specifications relating to normative requirements and references. These employees can be allocated in standardization or technical departments. In general, they are mainly very experienced employees with a strong technical background, (Drechsler, 2016).

2.1.5 Standard Developer

Within the literature the term standard developer has been introduced to describe a person, who represents a company in committees and who needs to be able to conduct multicultural, consensus-based negotiations, to listen, to observe and to estimate risks (McMillian, 2013).

Necessary knowledge and skills directly related to standardization of these groups are (Table 1).

In addition, they need to have a strong technical background, good language skills and a variety of soft skills, which are not described individually here (Blind & Drechsler, 2017).

2.2 Competence Demands of Individual Departments in Relation to Standardization

Aside employees, whose main task is the field of standardization, companies are represented in committees by employees who mainly work, for example, in technical departments. They are involved in standardization issues in their specific field

Table 1 Knowledge and skill profile standardization experts (Blind & Drechsler, 2017)

Necessary knowledge and skills in the field of standardization
− National, European and International Std. Committee Landscape
− Standardization and Legislation
− Micro- and macro-economic benefits of standards
− Standards and Patents
− Standardization process including rights and obligations in committees
− Strategic aspects of a participation in standardization
− Ability to write a standard
− Standard implementation
− Standard, conformity assessment and certification

of activity as they are the experts for a certain technical matter. Demands of the industry are normally directly related to a particular job description, e.g. development engineer, sales employee or production manager. As a result, the following literature review focuses on company departments and super-ordinate job-profiles such as for example business strategists. Initially, specific demand groups need to be identified as well as their requirements related to skills and competences in the field of standardization, (Blind & Drechsler, 2017).

The current state of research differs between employees applying standards and employees representing their companies in committees. However, knowledge; skills and competence requirements of company representatives in committees coincide with the demands of employees applying standards. In addition, employees involved in standardization are required to know the development process of standards, the rights and duties of committee work and to be able to write a standard (Drechsler, 2016).

Within the state of research demands in relation to standardization have been formulated for the different departments in companies of the private industry, e.g. IFAN, (2014).[1] These can be summarized as follows (Fig. 2).

An overview about their demands in relation to standardization of different departments is given in Table 2.

In addition, particular demands have been formulated for employees of SDOs and employees of regulatory and competition authorities, whereby these have received only marginal attention in research (Blind & Drechsler, 2017).

---

[1]A detailed analysis of the state of research on this topic can be found in the final report of the European study (Blind & Drechsler, 2017).

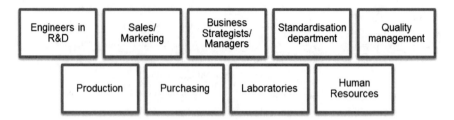

Fig. 2 Overview of demand groups in the state of research, classified by departments (Blind & Drechsler, 2017, p. 36)

3 Current Job Profiles and Their Demands in Europe

The above provided overview is based on the current state of research regarding particular demands in relation to standardization for different job profiles. A European study (Blind & Drechsler, 2017) on this topic deals among other things with the question, which standardization-related job profiles are currently required by industry. In addition to a detailed literature study, an empirical study is carried out which is divided into 3 studies[2]:

(1) **Analysis of job offers**: 300 job offers across Europe, which are directly related to external standards and standardization, have been analysed regarding the following information: origin of job offers (country and sector), job title and requirements on applicants. (see Sect. 3.1)

(2) **Analysis of LinkedIn profiles**: 270 LinkedIn profiles of experts having "standardization" in their profile (either in the description of their competences or their job profiles) have been analysed delivering information about the current/former position, field of study as well as main competences of an employee (see Sect. 3.2).

These two studies have revealed the sectors standardization experts are active in, their educational background as well as their knowledge, skills and competences. However, these approaches are not able to reveal in a systematic and detailed way the competences employees involved in standardization need for their job. Therefore, a survey has been performed specifically addressing these questions.

(3) **Survey**: The link to the survey has been distributed among our relevant LinkedIn contacts, the LinkedIn Groups IEC professional, World of Standards, ETSI people, Education about standardization, ICES, SMEs and Standards and the participants of the workshop hosted by the European Commission on the 16th of May 2017 asking to distribute the survey among their industry contacts. A total of 451 persons have participated in the study, whereof 193 respondents have completely filled out the questionnaire. The questionnaire itself addressed

---

[2] More detailed information about the research design can be found in the final report of the European study, cf. (Blind & Drechsler, 2017).

Table 2 Overview about necessary knowledge and skills (companies perspective) (Blind & Drechsler, 2017, p. 38)

Number	Competence	Engineers in R&D	Sales/marketing	Marketing	Business strategists/managers	Standardization department	Quality management	Production	Purchasing	Laboratories	Human resources
1	To know basic terms in the fields of standardisation	x	x	x	x	x	x	x	x	x	
2	To know basic standards and guidelines in one's field of expertise	x	x	x	x	x	x	x		x	
3	To know general basic standards	x			x	x	x	x		x	
4	To understand general relevance of standards from the economies and companies point of view	x	x	x	x	x	x	x	x	x	x
5	To know the international committee landscape	x		x	x	x	x				
6	To be able to research standards	x		x	x	x	x	x	x	x	
7	To be able to screen a standard	x				x	x	x		x	
8	To be able to estimate the consequences of an omission of the application of relevant standards in certain application	x				x	x	x		x	

(continued)

Table 2 (continued)

Number	Competence	Engineers in R&D	Sales/marketing	Marketing	Business strategists/managers	Standardization department	Quality management	Production	Purchasing	Laboratories	Human resources
9	To be able to choose relevant standards for a specific application	x				x	x	x		x	
10	To be able to interpret a standard	x				x	x	x		x	
11	To be able to take into account relevant contents of a given standards for the development of a product or a process	x				x	x	x		x	
12	To be able to evaluate if a product or process meets the required standards in one's area of expertise	x				x	x	x		x	
13	To know the formation process of standards	x		x	x	x	x				
14	To be able to create a standard					x					
15	To be able to estimate consequences of a new standard or specific content of a standard	x				x					

(continued)

Table 2 (continued)

Number	Competence	Engineers in R&D	Sales/marketing	Marketing	Business strategists/managers	Standardization department	Quality management	Production	Purchasing	Laboratories	Human resources
16	To know and understand the macroeconomic benefit	x	x	x	x	x					
17	To understand strategic aspects of a participation in standardisation	x	x	x	x	x	x			x	
18	To estimate potential possibilities of influence from a company's point of view	x			x	x					
19	To estimate the participation in strategic relevant consortia					x					
20	To initiate new standardisation issues					x					
21	To know rights and obligations in consortia					x					
22	To know and understand the role of standards in management systems and company management system policies			x	x						

(continued)

Table 2 (continued)

Number	Competence	Engineers in R&D	Sales/marketing	Marketing	Business strategists/managers	Standardization department	Quality management	Production	Purchasing	Laboratories	Human resources
23	To know and understand the interaction between innovation, intellectual property and standards (a necessary competence of employees in R&D	x									
24	To be able to identify conflicts with patents and standards or gaps in existing standards	x									

Necessary Competences of Employees in the Field … 125

among other things the career of the participant, the current working place and job, the most relevant skills and competences in the area of standardization for the current position, trainings on the job in the area of standardization and standardization contents in the own field of study (see Sect. 3.3).

In addition, identified knowledge, skills and competences in the area of standardization have been evaluated from the participant's point of view. Therefore, the participant has been asked to evaluate the knowledge, skills and competences regarding the requirement of the current position (see Sect. 3.4).

The following sub-chapters present the results of the different sub—studies.

3.1 Current Job Offers with Relation to Standardization

In order to identify current positions in the field of standardization, job advertisements related to standardization were looked for and analysed across Europe using the current job platforms step stone, LinkedIn and monsters. The focus was on jobs related to external standardization, while jobs referring to internal standardization of workflows and processes will be excluded of the study, because they are in general disconnected from any type of external standardization, which was the focus of the study. (Blind & Drechsler, 2017)

An analysis of 300 job offers across Europe show that there are no standardised job descriptions available in the field of standardization as they heavily vary. Searched positions can only be classified into jobs with (Blind & Drechsler, 2017):

- **Technical title**: Jobs with focus on technical tasks and background, e.g. Engineer standardization/ CE-Management
- **Business title**: Jobs with focus on business tasks and background, e.g. Finance Analyst
- **Programmer/ IT title**: Jobs with focus on IT tasks and background, e.g. Developer Java
- **Else**: All job descriptions, that do not fit into one of the other clusters.

However, it has to be noted that available job offers in the field of standardization are mostly technical jobs which are active in the sectors shown in (Fig. 3):

As available job offers heavily differ, not only in their description, but also in their actual content, expected hard skills also vary widely. Hence, it is not possible to make a statement about generally required skills. They can only be summarized three super ordinate categories (Blind & Drechsler, 2017)

- Technical knowledge and skills, e.g. Fluid construction, 3D CAD, Automation
- Business knowledge and skills, e.g. Lean Management
- IT- knowledge and skills, e.g. Software development, C#, Knowledge of particular software packages.

Aside the different mentioned hard skills, companies are looking for a certain soft skill profile of an applicant. The analysis of the job adds show, that the most

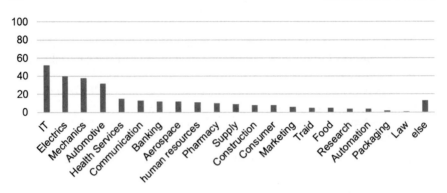

Fig. 3 Sectors with a demand of employees in standardization (Blind & Drechsler, 2017, p. 46)

important soft skills are team work (16%), communication ability (14%), personal initiative (10%) and management (7%) (Blind & Drechsler, 2017, p. 49). A study based on a sample of 297 German companies regarding the importance of certain soft skills reveals that the five most important soft skills are personal initiative, analytical thinking, communication ability, team work and willingness to perform (Praktikum Info, kein Datum). The comparison with the findings of (Blind & Drechsler, 2017) shows that required soft skills are quite similar, which means they do not appear particularly special for activities in the field of standardization.

Regarding the level of education, most companies are looking for applicants holding a university degree, whereby only 4% are addressing foremen in their job offers and only 1% PhDs. But what is more important, 71% of the job offers require professional experience in general, even though only 4% explicitly demand experience in the field of standardization (Blind & Drechsler, 2017, p. 48).

3.2 Current Job Profiles in Relation to Standardization

A total of 270 LinkedIn profiles of standardization professionals having "standardization" in their profile (either in the description of their competences or their job profiles) have been analysed across Europe in order to identify the most relevant sectors and countries. In addition, such a profile analysis delivers information about the most relevant field of studies, the competences and their correlation with other competences (Blind & Drechsler, 2017).

The majority of employees are based in Germany. First, Germany is the strongest economy in Europe. Second, Germany is the world leader in standardization according to International Organization for Standardization ISO (ISO, 2017), which explains the high number of employees in this particular field.

Most employees working in standardization hold a Master/ Diploma degree (66%) or even a PhD (16%), whereby the percentage of employees holding a Bachelors only amount to 9%, which is relatively low, (Blind & Drechsler, 2017). The field of study can be differentiated into 5 main clusters (Blind & Drechsler, 2017, p. 43):

- **Engineering (55%)**: This cluster contains all technical fields with reference to engineering, e.g. mechanical engineering, electrical engineering, civil engineering, aeronautical engineering, production engineering or chemical engineering.
- **Management (10%)**: This cluster contains all fields of study with reference to management and business, e.g. Economics, Leadership& Management, general management, business or business administration.
- **Sciences (5%)**: This cluster contains all fields of study with reference to natural sciences, e.g. physics, chemistry or mathematics.
- **IT (18%)**: This cluster contains all fields of study with reference to telecommunication, computer sciences and programming.
- **Else (12%)**: This cluster contains a mixture of different fields of study, which are more unusual, e.g. law and international relation.

The fact that 40% of all employees holding a degree in management have studied this subject as a second degree is surprising, (Blind & Drechsler, 2017). From this, it can be concluded that a combination of technical and management skills is required in the field of standardization.

The job offer analysis shows that most companies require professional experience when hiring employees for activities related to standardization. The results of the LinkedIn analysis support these finding by showing that most employees have professional experience before starting working in standardization, cf. Figure 4. (Blind & Drechsler, 2017).

That means that employees are changing the field of activity towards standardization later on in their career.

The study focuses on the main five knowledge and skills of a member. The results show that the most important competences are project management/-planning

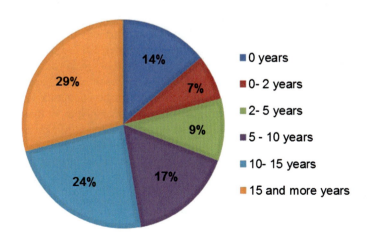

Fig. 4 Time before working in standardization (Blind & Drechsler, 2017, p. 44)

(19%) and business strategy (10%)—both are non-subject specific competences. They are followed by subject specific competences such as knowledge about 3GPP[3] (9,2%), Telecommunication (8,5%) and Wireless (8,5%). Hence it is notable that the most common competences of employees in standardization are competences in IT/Telecommunication (Blind & Drechsler, 2017, p. 45).

3.3 Current Job Profiles in the Field of Standardization—Results of a Survey

This section describes the characteristic of the respondents of the survey regarding sector, company size, department, degree and field of study, working experience and current activities in standardization.

The respondents are very active in standardization. On the one hand, two thirds are active in external standardization bodies, particularly at the national level. However, half of the respondents are either involved in European or international standardization. Finally, more than 20% are already active in consortia and for a confirming of trend already observed within the German Standardization Panel (Blind & Müller, 2016). Consequently, it can be assumed that the results of the survey with regard to the necessary competences provide a representative picture of actual needs, (Blind & Drechsler, 2017).

One important aspect is the size of the organisation. Almost half of the respondents work in organisations with more than 500 employees. However, one quarter is working in organisations up to 49 employees, (Blind & Drechsler, 2017, p. 55).

A differentiation of the respondents by sectors reveals the important role of electro technical engineering (26%), but also the rising importance of services (16%). Almost 10% of the respondents are involved in research and education, which was admittedly not the target group of the survey. Furthermore, a similar number of participants work in construction, manufacturing, IT and mechanical engineering. Finally, experts working in standardization, certification and metrology institutions and public administration complete the sample (Blind & Drechsler, 2017, p. 55). The result is also reflected in the analysis of the job offers, since mainly specialists are sought for a specific department or activity, e.g. R&D. Standardization is just one part of their professional activities. The survey also provides information about the departments in which the respective employees are employed. Surprisingly, more than one third of the respondents work in the R&D department, despite the limited coordination with research and development revealed in other studies, e.g. (Großmann, Filipovic & Lazina, 2016). In addition, 15% are active on the general management level, 13%

---

[3]"The 3rd Generation Partnership Project (3GPP) unites seven telecommunications standard development organizations […] The project covers cellular telecommunications network technologies, including radio access, the core transport network, and service capabilities—including work on codecs, security, quality of service—and thus provides complete system specifications." (3GPP Mobile Competence Centre, 2017).

in quality management and less than 10% in the standardization department. Overall, only a small fraction of the respondents with an interest in standardization is active at the general management level, which is challenging the general promotion of the topic within companies, whereas we find a large share active in R&D, which might be an interesting opportunity for its future promotion, (Blind & Drechsler, 2017, p. 56).

More than half of the respondents hold a Master's degree, whereas almost one quarter holds a Bachelor or a PhD. Most degrees have been awarded by universities (Blind & Drechsler, 2017, p. 57). This finding is confirmed by the job offer analysis showing that 80% of the companies expect applicants having a university degree, whereby it is in general not further specified. Only 4% of the job offers are addressing foremen and only 1% requires a PhD as a precondition (Blind & Drechsler, 2017, p. 48). The scientific disciplines are dominated either by engineering in general (39%) or IT engineering (31%), incl. computer sciences, telecommunications, etc. These two thirds having either an engineering or IT background are confirmed both by the studies based on LinkedIn profiles and on job profiles. The remaining third is divided into experts with a natural science (14%) or a social science (16%) degree. Overall, the distribution of the scientific background of the respondents to the survey corresponds quite well to the composition of the LinkedIn job profiles and the job titles, which supports the validity both of our approach and the collected data (Blind & Drechsler, 2017).

Finally, the relevance of standardization in the professional activities of the respondents is confirmed by the majority being confronted with standardization already in their job interviews, whereas for two thirds standardization has explicitly been added to their job responsibilities, (Blind & Drechsler, 2017).

Whereas previous experience supported the perception that the career of standardization professionals starts only after working some time in other (technical) areas, the survey reveals a slightly different pattern. More than one third of the respondents reports that standardization belonged to their job responsibilities within the first year of employment and for almost another one third within the second to the fifth year. Only for 27.5% standardization is added to their job responsibilities after six years of employment, (Blind & Drechsler, 2017).

3.4 Knowledge, Skill and Competence Demands of Employees in Standardization

The approaches presented above allow deriving job profiles as well as personal profiles of employees working in standardization. However, they are not able to reveal in a systematic and detailed way the competences they need for their job.

Based on the competency model developed by Drechsler, (2016) focusing on the machinery sector, the insights from the literature review and some additional interviews, a list of knowledge, skills and competences needed by employees either implementing standards or participating in standardization processes has been developed. The competences are expected to build on each other, i.e. we start with the basic competences followed by the more sophisticated ones. Within the survey all participants have been asked to evaluate the following knowledge, skills and competences regarding the requirement of the current position

1. To know the basic terms used in standards and standardization
2. To be able to identify the need for standards
3. To be able to search for and select appropriate standards
4. To be able to implement standards in product or process development
5. To be able understand the impact of implementing standards
6. To know the standardization institutions and their processes
7. To be able to identify the need for getting involved in standardization
8. To be able to search for and select appropriate standardization organisations
9. To be able to be passively involved in standardization processes (observer)
10. To be able to be actively involved in standardization processes (participant)
11. To be able to judge which form of standardization (formal vs. consortia) is appropriate
12. To be able to propose new work items in standardization
13. To be able to strategically influence the agenda in standardization processes
14. To be able to understand the impact of standardization processes.

The evaluation has been carried out using a scale ranging from "not relevant", "useful", and "very relevant" to "indispensable to assess the relevance of competences related to standards and standardization for handling work task. In addition the participants has been asked for the sufficient or insufficient provision of competences by higher or other education institutions, but also whether this is not necessary.

The results of the survey are shown in Fig. 5.

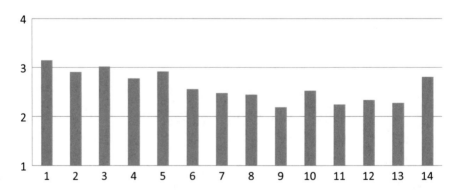

Fig. 5 Relevance of competences (1 = not relevant to 4 = indispensable) (N $\geq$ 202) (Blind & Drechsler, 2017, p. 61).

3.5 Evaluation of knowledge, skills and competencies

The answers displayed in Fig. 5 reveal in general that the proposed hierarchical competence model composed of the 14 specific competences is confirmed by the ranking of competences needed in practice. However, there are a few exceptions noteworthy to be mentioned. More important than according to the hierarchical competence model expected are understanding both the impacts of implementing standards and of being involved in standardization processes, (Blind & Drechsler, 2017, p. 60).

Aside the evaluation of the knowledge, skills and competences regarding the requirement of the current position, the participants were asked to evaluate if a sufficient or insufficient provision of competences has been taken place by higher or other education institutions, but also whether this is not necessary. The results are illustrated in Fig. 6.

It is notable, that a more or less mirror image of the ranking of the relevance of competences is the percentage of the respondents indicating that the competences gained at higher or other education institutions are insufficient in handling their work tasks, i.e. 0 means. In general, most respondents did not receive sufficient competences related to standards and standardization in their education. This pattern is also a validation of the ranking of the relevant competences. (Blind & Drechsler, 2017, p. 59)

In addition to the list of presented competences, more than thirty respondents proposed further competences. On the one hand, the interactions between standards and the regulatory framework are mentioned several times. In addition, a better understanding of standardization in the context of the national quality infrastructure and the macroeconomic environment is requested. On the other hand, soft skills, like related to negotiation, diplomacy and networking are required, (Blind & Drechsler, 2017, pp. 59–60).

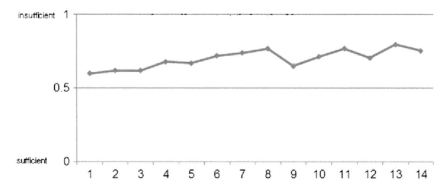

Fig. 6 Evaluation of the question « To what extent are the competences gained at higher or other education institutions insufficient in handling tasks? » (N = 128) (Blind & Drechsler, 2017, p. 61)

4 A Competence Model for Standardization

The elementary abilities and their connections in a hierarchical way are described in competence models by means of competence dimensions, whose number varies depending on the particular field. The competences itself can be independent of each other or hierarchical related to each other, depending on the field. General frameworks for generic competence models provide for example the Bloom's taxonomy (Krathwohl, 2002), which creates a common language between and among educators. It provides a tool to define objectives in order to create a curriculum by using the following hierarchy within cognitive process dimensions: remember, understand, apply, analyze, evaluate and create (Conkling, 2005).

Even though standardization is a cross-circular theme, it has got specific demands for abilities. That means detailed cognitive processes from generic models have to be consolidated into relevant competence dimensions for the specific field of standardization. This leads to a stronger focus on cognitive processes (e.g. factual or procedural knowledge) within the particular domain specific competence model (Albers, et al., 2015).

First it is necessary to define core abilities and demands in selected competence areas to develop a competence model, which can take place by several means, including interviews, observations, questionnaire surveys and research investigations. The competences itself need to be defined in a way they represent the theoretically and empirically verifiable basis for the development of test tasks in order to verify the model (Albers, et al., 2015).

As the answers displayed in Fig. 5 confirm that the proposed hierarchical competence model composed of the 14 specific competences meet the demands of the industry, they form the basis for the following competence model. In a next step, these will be extended by the competences proposed by the survey respondents (Blind & Drechsler, 2017):

15. To be able to understand the interactions between standards and the regulatory framework
16. To be able to understand standardization in the context of the national quality infrastructure and the macroeconomic environment
17. To know the history of standardization.

The aim of the following competence model is to describe the elementary abilities and their connections in a hierarchical way. This means that the individual competences have to be put in relation to each other and arranged hierarchically.

The base is built by the competence *"To know the basic terms used in standards and standardization"*, which knowledge is a basic prerequisite to be able to acquire further competences in this particular field. As an example, a student needs to know the definition of a standard as well as the difference between standards and guidelines in order to be able to understand the impact of standards for a company. The same procedure can be applied to each competence which leads to the result shown in Fig. 7 and therefore provide a base for education in standardization (Blind & Drechsler, 2017).

Necessary Competences of Employees in the Field …

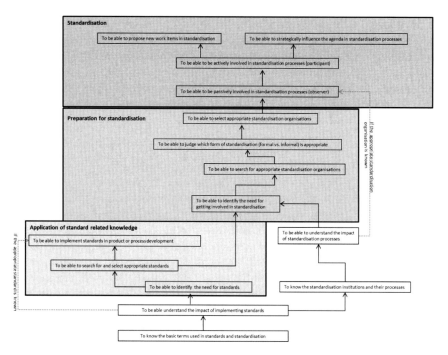

Fig. 7 General competence model for the field of standardization (Blind & Drechsler, 2017, p. 77)

It should be noted that all competences (cf. Figure 7) are hard skills. As already mentioned several times, soft skills also play a decisive role in the field of standardization, e.g. negotiation, diplomacy, networking, communication, team work and self- responsibility. These are not part of the competency model, since these are cross-disciplinary qualifications that cannot be assigned exclusively to the area of standardization. Rather, these are specific traits of a person or qualifications that can and should be acquired in a different context. The individual competences can be divided into four different clusters (Blind & Drechsler, 2017):

(1) **General competences**: Competences representing general knowledge, such as for example "to know the basic terms used in standards and standardization"
(2) **Application of standard related knowledge**: Competences which are required in order to apply a standard
(3) **Preparation for standardization**: Competences which are required to prepare standardization activities of a company
(4) **Standardization**: Competences which are required for an active participation in standardization.

These categories are not independent of each other, they build upon each other. This means that an employee has to master the knowledge of the subordinate category in order to be able to dispose of the superior competences. This differentiation can be used for an alignment of higher education in the field of standardization as not all knowledge, skills and competences can be taught at universities and need to be acquired in trainings on the job or even by professional experience (Blind & Drechsler, 2017). Providing the necessary skills is another major challenge, as this can only be achieved through a combination of different educational pathways. Among other things, the different education systems of the various European countries as well as company internal training must be considered. Accordingly, this topic must be discussed in depth and in detail afterwards. Consequently, it is not part of this article because its vastness.

5 Summary and Conclusion

In general, the degrees and field of studies revealed by the job profiles, the job offers and the responses to the survey are consistent and differ only slightly due to different categories and classification used. Although the literature review indicates some focus on engineering the quantification, we are able—based on our three empirical approaches to confirm this previous finding, but also to specify it further towards electro technical and information technology. Consequently, the majority of the experts are also working in these or related sectors. All other areas are of minor importance corresponding to observations from the literature review, (Blind & Drechsler, 2017, p. 72).

The business unit, the experts are mainly working in, is as a further consequence also research and development, which have a high relevance in engineering. However, we find professionals also on the general management level or quality management in particular, which is not contradicting the findings from the literature, (Blind & Drechsler, 2017, p. 72).

According to the LinkedIn analysis and the results of the survey most experts are holding a Masters degree awarded at universities, which is confirmed by the job offer analyses. However, the results of the survey show that necessary competences in the area of standardization are not sufficiently taught at universities and other higher education institutes, (Blind & Drechsler, 2017, p. 72).

Companies are looking for applicants with professional experience before starting to work in a position with relation to standardization. This observation is supported by the results of the LinkedIn analysis, according to which more than half of the experts start working in standardization after more than 10 years on the job. The results of the survey differ from those of the LinkedIn analysis, which can be explained by the fact that the average age is higher for the LinkedIn group, (Blind & Drechsler, 2017, p. 72).

Furthermore, the analysis of the job offers, and the LinkedIn profiles reveals numerous technology or even standard-related contents, which are sector-specific, e.g. telecommunication or ICT. In addition, necessary soft skills mentioned in the literature are also found in the descriptions of the job offers. However, the multi methodology approach of the study delivers 17 standardization specific competences, which meet the demand of the industry. These competences are transferred into a hierarchical competence model showing on the one hand the relations between each other and on the other hand providing a base for education in standardization. How this is then implemented in the future has to be examined and discussed in detail in further research, (Blind & Drechsler, 2017, p. 72).

Overall, on the one hand the consistency of our findings within the various approaches applied in our study is surprisingly high. On the other hand, there are no significant differences to the insights from previous studies. However, our quantitative approach allows us some further differentiation, which questions the differences between disciplines and business units found in the literature. In addition, we are able to identify some technology or even standard-specific requirements for knowledge, skills and competences, which cannot be found in the existing literature, (Blind & Drechsler, 2017, p. 72).

References

3GPP Mobile Competence Centre. (2017). *3GPP—A global initiative*. [Online] Available at: http://www.3gpp.org/about-3gpp.

Albers, A., Butenko, V., Breitschuh, J., Walter, B., Drechsler, S., & Burkardt, N. (2015). Guidelines for competence assessment in engineering education an impementation in project NuSaL. In ICED, Hrsg. *Proceedings of the 20th International Conference on Engineering Design*. Milan Italy: s.n.

Bailetti, A. J., & Callahan, J. R. (1995). Managing consistency between product development and public standards evolution. *Research Policy, 24,* 913–931.

Blind, K., & Drechsler, S. (2017). *European Market Needs for Education in Standardisation/Standardisation-related Competence*. Brüssel: European Commission.

Blind, K. & Mangeldorf, A. (2016). Motives to standardize: Empirical evidence from Germany. *Technovation*, February–March, pp. 13–24.

Blind, K., & Müller, J.-A. (2017). *Indikatorenbericht 2017, Deutsches Normungspanel, Normungsforschung, -politik und -förderung*. Berlin: Deutscher Förderverein zur Stärkung der Forschung zur Normung und Standardisierung e.V.

Blind, K., & Müller, J.-A. (2016). FNS und TU Berlin stellen die Ergebnisse der vierten Befragungsrunde des Deutschen Normungspanels (DNP) mit Fokus auf Digitalisierung vor. *DIN-Mitteilungen, 07,* 21–25.

Carugi, M. (2013). *Industry needs session 1: Company needs for standardisation education*. Sophia Antipolis, ICES conference and meeting.

Conkling, J. (2005). Book Reviews: A Taxonomy for learning, teaching and assessing. In: *A revision of Bloom's Taxonomy of Educational Objectives*. s.l.: Educational Reviews, pp. 154159.

Cooklev, T. & Bartleson, K. (2008). Institute of Electrical and Electronics Engineers—The strategic value of standards education. In: D. E. Purell, Hrsg. *The Strategic Value of Standards Education*. Washington D.C.: The Center of Global Standards Analysis, pp. 34–37.

Drechsler, S. (2016). *Competency Requirements implied by Standards and Guidelines for Mechanical Engineers in their day-to-day Activities; Ph.D. Thesis,* Karlsruhe: IPEK.

Drechsler, S. & Albers, A. (2016). A competency model for standardisation in the field of standardisation. In I. Horvath, J. Pernot & Z. Rusak, Hrsg. *Proceedings of TMCE 2016*. Aixen-Provence, France: TMCE.

European Commission. (2011). A strategic vision for European standards: Moving forward to enhance and accelerate the sustainable growth of the european economy by 2020. In *Communi-*

cation from the commission to the european parliament, the council and the european economic and social committee, s.l.:COM (2011) 311 final.

European Commission. (2016). European Standards for the 21st century. In *Communication from the commission to the European Parliament, the council, the European economic and social committee of the regions.* Brussels: COM (2016), p. 358.

Freericks, C. (2013). *Industry needs for standards engineers—Results from a global inventory.* ICES: Sophia Antipolis.

Großmann, A.-M., Filipovic, E., & Lazina, L. (2016). The strategic use of patents and standards for new product development knowledge transfer. *R&D Management, 46*(2), 312–325.

Hesser, W. & de Vries, H. J. (2011). *Academic Standardisation Education in Europe,* Hamburg, Rotterdam: Euras—European Academy for Standardisation e.V.

Hövel, A., & Schacht, M. (2013). Standardisation in education—A building block for professional careers. In H. de Vries & H. Brand (Eds.), *2013 ICES Conference, What does Industry expect from education about standardisation? Proceedings* (pp. 6–15). ICES: Sophia Antipolis.

ISO. (2017). *Number of secretariats listed by the technical committees and subcommittees of the International Organisation for Standardisation (ISO).* [Online]. Available at: www.iso.org/iso/home/about/iso_members.htm.

ISO. (2014). *Teaching standards; Good practices for collaboration between National Standards Bodies and universities,* Geneve: ISO Central Secretariat.

IFAN. (2014). *Education and training about standardization, different needs for different roles, guide 4:2014.* Geneve, Switzerland: IFAN.

Jakobs, K., Procter, R. & Williams, R., (2001). The Making of Standards: Looking inside the Work Groups. *IEEE Magazine,* 04.

Krathwohl, D. R. (2002). A revision of bloom's taxonomy: An overview. *Theory into Practice, 41*(4), 212–218.

Kurokawa, T. (2005). Developing human resources for international standards. *Science & Technology Trends—Quarterly Review,* pp. 34–47.

McMillian, A. (2013). *A Rockwell Automation Perspective, ICES 2013 Conference—What does Industry expect from Standards Education?* Sophia Antipolis, ICES.

Nagel, H. (2002). Strategische Bedeutung von Normen für klein- und mittelständische Unternehmen (KMU) im globalen Wettbewerb. In T. Bahke, U. Blum, & G. Eickhoff (Eds.), *Normen und Wettbewerb* (pp. 67–77). Berlin: Beuth Verlag.

Praktikum Info, kein Datum *Was sind Soft Skills? Eine Liste zum Überblick.* [Online]. Available at: https://www.praktikum.info/karrieremagazin/bewerbung/soft-skills [Zugriff am 18 08 2017].

Tanaka, M. (2010). *Education needs and JSA activities.* Geneve: WSC Academic week.

The Center for Global Standards Analysis. (2008). *The strategic value of standards education.* Washington D.C.: Donald E. Purcell, Chair.

Weiss, M. B. (1993). The standards development process: A view from political theory. *StandardView, 1*(2), 12, 35–41.

Knut Blind is Professor of Innovation Economics at the Technische Universität Berlin. He studied economics, political science and psychology at Freiburg University. In the course of his studies he spent one year at Brock University (Canada), where he was awarded a BA. Finally, he took his Diploma in Economics and later his doctoral degree at Freiburg University. Between 1996 and 2010 he joined the Fraunhofer Institute for Systems and Innovation Research, Karlsruhe, Germany, as a senior researcher and at last as head of the Competence Center "Regulation and Innovation". In April 2006 Knut was appointed Professor of Innovation Economics at the Faculty of Economics and Management at the Technische Universität Berlin. Between 2008 and 2016 he held also the endowed chair of standardisation at the Rotterdam School of Management, Erasmus University. From April 2010 he is linked to the Fraunhofer Institute of Open Communication Systems in Berlin. In 2012, he initiated both the Berlin Innovation Panel and the German Standardization

Panel. He published several articles on various aspects of innovation economics, in particular contributions on regulation and standardization, but also Intellectual Property Rights.

Sandra Drechsler has been working since October 2017 as a consultant for technology and standardization at the Association of German Machine and Plant Manufacturers and supports numerous international and European standardization committees in the field of metallurgical and rolling mill equipment as well as foundry machinery. First, she studied Mechatronics at University of Cooperative Education, Mannheim, and afterwards Mechanical Engineering at the Karlsruhe Institute of Technology (KIT). In addition to her studies at KIT, she worked as a mechanical design engineer at Schenck Process GmbH, Darmstadt, from 2006 to 2009. After completing her mechanical engineering studies in August 2010, she began working as a research assistant at the Chair of Product Development at KIT and received her doctorate in June 2016 with the topic "Competency Requirements implied by Standards and Guidelines for Mechanical Engineers in their day-to-day Activities", which was honoured by DIN with the special scientific award. During her doctorate she oversaw among other things the BMBF project "Standardization in Academic Education" in cooperation with DIN from 2013 to 2015. After completing her doctorate, she conducted in 2017 the European service study "European Market Needs for Education in Standardisation/Standardisation-related Competence" together with Prof. Blind/Fraunhofer Fokus.

UNECE Initiatives on Education on Standardization

Lorenza Jachia, Serguei Kouzmine and Haiying Xu

1 Introduction to UNECE

In its area of competence, the United Nations Economic Commission for Europe (UNECE) plays a vital advocacy role for the use of standards as part of regulatory frameworks and policy work. Headquartered in Geneva, at the Palais des Nations, UNECE is an intergovernmental organization, one of the five UN regional economic and social commissions.

UNECE pursues pan-European economic integration, and helps implement the 2030 Agenda on Sustainable Development and other landmark United Nations (UN) frameworks by translating global goals into norms, standards and conventions. Since the 56 member states of UNECE, which include European, North American and Asian countries, account for around half of global trade[1]; many of UNECE standards have global significance. As such, all interested UN Member States may participate as observers—and in some cases with full voting rights—in the "rulemaking" work

---

L. Jachia (✉)
UNECE Working Party on Trade and Standardization Policies, Economic Cooperation and Trade Division, UNECE, Geneva, Switzerland
e-mail: lorenza.jachia@un.org

S. Kouzmine
UNECE, Geneva, Switzerland
e-mail: serge.kouzmin@gmail.com

H. Xu
Intern, Economic Cooperation and Trade Division, UNECE, Geneva, Switzerland
e-mail: haiying.xu@graduateinstitute.ch

[1]In terms of merchandise trade, data is available for 52 countries out of the 56 UNECE member countries in the statistics database of the World Trade Organization (http://stat.wto.org/Home/WSDBHome.aspx); their aggregate exports account for 51.82% of world's total exports, and their aggregate imports 54.84% of world's total imports.

© Springer Nature Switzerland AG 2020
S. O. Idowu et al. (eds.), *Sustainable Development*, CSR, Sustainability, Ethics & Governance, https://doi.org/10.1007/978-3-030-28715-3_9

of UNECE. UNECE is also open to the participation of external organizations: over 70 international organizations and other non-governmental organizations regularly take part in UNECE activities.[2]

In addition to its normative activities, UNECE actively cooperates with many standards-setting organizations. As one example, in the area of electronic business, UNECE, International Organization for Standardization (ISO), International Electrotechnical Commission (IEC) and International Telecommunication Union (ITU) are signatories of a joint Memorandum of Understanding on electronic business.[3] In many other fields, UNECE intergovernmental bodies and groups of experts work closely with relevant international and regional standards setting organizations; for example, in the sector of transport, construction of vehicles, information exchange, etc.

Furthermore, UNECE promotes the use of international standards, whenever relevant, as a means of facilitating trade, harmonizing national technical and regulatory requirements, and translating international, regional and national policy goals into action on the ground.

2 Role of Standards in a Globalized World

Standards have a pervasive impact on all aspects of our daily lives, as they lay out the characteristics of products and services we all use. While we take it for granted that a sheet of paper will slide through the photocopy machine and that our credit card will enable us to pay for purchases in any country, these and many other routine activities would be simply unimaginable without international standards. International trade crucially depends on international standards which enable buyers and sellers to agree on desired contract terms without ambiguity.

Allowing the consistent monitoring and improvement of the quality and reliability of production processes, standards are essential to business. Additionally, they are important tools for managing risks that confront consumers, workers, and communities, especially those who live close to dangerous production plants or in zones exposed to natural hazards.

In many ways, standards are also facilitating the transition to a green economy and to more sustainable production and consumption patterns, enabling companies to reduce their energy consumption and carbon footprint, and to ensure that resources are used responsibly and are preserved for future generations.

Coming to the subject matter of this paper, successfully tapping the potential of standards, both at the individual and at the societal level, requires at least a basic

---

[2]For more information please visit UNECE website (http://www.unece.org); in particular "International Regulatory Cooperation: The Case of the United Nations Economic Commission for Europe" by UNECE and the Organisation for Economic Co-operation and Development (OECD). Available at: https://www.unece.org/index.php?id=47668&L=0.

[3]Memorandum of Understanding on electronic business, UNECE, ISO, IEC and ITU. Available at: http://www.itu.int/ITU-T/e-business/files/mou.pdf.

awareness and understanding of standardization. For example, to access international markets and integrate global supply chains, businesses need to become familiar with the requirements of one or several standards to be able to meet the desired product requirements. Additionally, managers need to be able to provide assurance of the high and stable quality of products and possibly also of production processes. Providing such assurance—that is being able to prove that standards' requirements have been met—may prove particularly challenging. This may include having the product tested, production facilities audited, and the impact of production processes on the environment verified. These activities can be well beyond the reach of many small and medium-sized enterprises.

Market developments are pushing towards increased products complexity, and a very short time-to-market. Products typically cut across different categories requiring different sets of standards to be adhered to simultaneously, adding to the challenge. The list of regulatory requirements for products is constantly expanding, including environmental standards, labor laws etc., thus requiring a company to fulfil an unprecedented number of obligations.

Standards are also used in many ways in the regulatory systems of important markets, in particular to promote public policy goals, such as curbing carbon emissions in the atmosphere, making buildings resilient to hazards, enhancing the compatibility of products and services, among many more. At the same time, regulatory measures may stem from protectionist motives. For instance, regulators may impose more restrictions on foreign products, request that a product is tested only in specific laboratories, or bases his requirements on unwarranted risks without scientific justification, etc. In 2012, the World Trade Organization (WTO) Trade Report focused on non-tariff measures which include standards. The report noted with concern that the increased application of technical barriers to trade posed a major challenge for international trade; the impact of non-tariff measures is difficult to quantify, yet there was evidence that they "significantly" distorted trade, "possibly even more than tariffs" (WTO, 2012, p.221).

Standards and regulatory requirements may be as much a contribution to enhanced quality and competitiveness as a hurdle to a firm's profitability and its ability to engage in international trade. Many of these perceptions—while in part true—may also in part reflect an insufficient awareness and understanding of standards and related requirements, as well as an insufficient capacity to participate in standards-setting activities. Hence, it is of great importance for policy makers and for businesses to be familiar with both the trade-facilitating and trade-distorting aspects of standards.

Increased understanding and use of voluntary standards, at a societal level, will minimize technical barriers to trade and strike an optimal balance between desired policy outcomes and their costs to society. In addition to static gains, this understanding will also foster positive dynamics at systems level. Indeed, a complex link exists between standards and innovations. On the one hand, development of standards can facilitate the introduction and implementation of innovations and their interoperability with the currently used devices and products. On the other, existing standards and technologies can hinder the adoption of new competing technical approaches.

Again, policy makers need more than a cursory understanding of the challenges to make optimal choices for their constituencies.

The remainder of this paper, building on understanding the importance of standardization to meet current and upcoming challenges, will consider the situation of education on standards and standards-related issues. Then the efforts of UNECE to improve education about standards will be introduced, followed by directions for future work.

3 Market Demand for Competencies Related to Standards

One way of understanding the need for competencies and skills related to standards is an analysis of processes typically performed by businesses and the extent to which they are informed by standards.

The International Federation of Standards Users (IFAN), which unites major industrial companies, published a guide showing what knowledge of standards-related issues are required at different levels or departments in a company (see Table 1) (IFAN, 2014).

Table 1 Understanding standards—requirements for company employees

Knowledge / Functions	Standardization role	Standards identification process	Integration of the content for relevant standards	Take into account sustainable development	Integration of management systems	Product compliance		Standardization lobbying		Business intelligence in standardization
						Issues	Methods	Issues	Methods	
CEO & senior management	•••			••	•••	•••		•••	•	
Human resources	•		•••							
Marketing	•••	•	•		•••	••		••	••	••
Sales	•••									
R&D and Innovation		•	••	••	••	•••		••	••	•
Laboratories			••		••	•••	•••	••	••	•
Purchasing		••		••	••					•
Production										
Quality, Environment, Safety Social responsibility			••	••	••	•••				•
Standardization	•••	•••	••	••	••	••	••	••	••	••

••• Substantial knowledge   •• Good knowledge   • Some Understanding

Source "Education and Training about Standardization", IFAN, 2014

Table 1 shows how almost every employee in a company from top management to marketing or sales agents needs at least a certain understanding of standards. For example, according to IFAN, CEO and top management are expected to have:

a. "substantial knowledge" of: need to comply with standards and regulations; strategies for standardization work; role of business associations in developing standards; standards and company management systems;
b. "good knowledge" of: how standards can help sustainable development (in the context of company's strategy at world market place); and
c. "some understanding" of: how to influence the content of standards from the company's strategic perspective; how to identify standards and regulations that products have to comply with. Relevant skills for CEO and top management include: identification of risks and opportunities; global implementation process; and influencing lobbying process, among others (IFAN, 2014).

Different levels of knowledge are also required from other employees at different levels, for example, in standardization (evidently), production, sales and marketing departments. Marketing experts, in addition to similar knowledge of general issues as that required from CEO, should also understand the value of compliance with standards, and know the methodology of obtaining information on development of standards. They shall also possess adequate knowledge to identify relevant regulations, to monitor them, to map standards into company's operations, to understand how to participate in a standardization process, and to relate sustainable development to product marketing (IFAN, 2014).

Another function that could be added to the otherwise comprehensive table above concerns legal departments. Clearly, legal departments will need an understanding of the technical requirements that form the basis of legal documents. In case of a litigation, the utilization of standards can be used to demonstrate that the company satisfied safety or other mandatory governmental requirements (as set by standards). This may be of relevance in cases where two parties have conflicting views over quality or other aspects of a delivered product. Moreover, standards are increasingly used in court to decide on the degree of responsibility. Therefore, proof of conformity to a set of standards provides a form of legal protection.

Another interesting study on the relevance of standards qualifications on the job market was carried out in Germany in 2010–2012 by analyzing around one thousand job offers from recruiting companies looking for professionals with the words "standards" and "engineers" in the job title (University of Bremen & DFKI–German Research Centre for Artificial Intelligence, 2012). The job offers that were analyzed were those targeting engineers (or candidates with equivalent bachelor degree) and those that required additional experience in working with standards (sometimes with a specific standard). Out of all advertised positions with "engineer" in the title, standardization activities were mentioned in 30% of job descriptions. Even discounting for a possible confusion regarding using and/or understanding the term "standard", the numbers remain substantial.

The positions with "standard" in the title included the following main areas of duties: managers (39%), engineers (29%), analysts (16%), developers (12%), compliance experts (6%), and trainers (2%). When considering the job offers by industry sector, out of all offers of "standards professionals" the sector "manufacturing" accounted for 34%, followed by "healthcare and hospitality" (25%) and "information technology" (15%). For "standards engineers" the data was: "manufacturing" (55%), "information technology" (25%), and "healthcare and hospitality" (7%), etc. (University of Bremen & DFKI–German Research Centre for Artificial Intelligence, 2012).

Although more research is needed, a preliminary conclusion is that many positions within a typical company require a portfolio of standards-related competencies and skills, and these competencies are actively sought by companies on the job market.

4 Education on Standards

In its area of competence, UNECE has—for more than 40 years—advocated for standards as an important policy and technical tool. Recently, UNECE activities focused on promoting voluntary standards as an essential part of sustainable development and called for stepping up the use of standards in the implementation of the 2030 Agenda for Sustainable Development.

As mentioned in the introduction, UNECE sets norms and standards, for example, in the area of agricultural quality, electronic commerce and automotive vehicles, among others. It also hosts a permanent group of experts that develop best practices and recommendations in the areas of standardization, conformity assessment, technical regulations, enforcement, and related flanking issues. This group has been active since 1970 and today it is called UNECE Working Party on Regulatory Cooperation and Standardization Policies (Working Party 6: WP.6). Already in 1970, the Working Party noted the importance of teaching standardization in high schools, vocational schools, universities and graduate schools.

An increasing number of discussions at the WTO and at other fora on technical barriers to trade in the 90 s brought education issues once again to the UNECE agenda in the context of what knowledge and skills relating to standards are required today by the market.

Standards have always been, and will continue to be, in the curricula of engineers in all countries, which has been confirmed by the above-mentioned study on job offers in Germany. UNECE complements the activities of various standardization bodies by providing horizontal awareness raising, training and education materials, specifically for students of non-technical subjects with an emphasis on the multi-disciplinary character of standards and their value for policymaking and management. The work of UNECE on education about standards and standards-related issues takes place within a rich context, as many countries and institutions are simultaneously engaged in this field. The remainder of this section provides an overview of the situation in many regions. The data on the overall situation of education about standards, e.g. the total number of institutions providing related courses, of courses taught, and of students who have received such training, are not readily available. It is hoped that future studies can provide us with these data. The following overview is based on the currently available data and selected examples.

4.1 APEC Countries

According to a survey made by the Helmut Schmidt University (Hesser, 2014), in the countries of the Asian region, Korea, Japan, Indonesia and China are leading the work on education about standards and standards related issues. Many universities in these countries actively pursue education on standards and standards-related issues.

In Japan, 51 courses were taught in 32 educational establishments in 2012. The number of trained students reached 2100 in that year. These activities also featured the publication of textbooks. In Korea, in 2011 there were 81 programmes offered at 41 universities with a total number of 3880 students. Notably, the number of lecturers on standardization in Korea increased from 50 in 2005 to 294 in 2011. In Indonesia, the national standards body signed a memorandum of understanding with 30 higher education establishments and standardization was taught in 10 universities in 2012. The number of participating students rose from 17 in 2007 to 450 in 2012, with a cumulative total number of 1036 students trained on standardization curricula between 2007 and 2012. In China, courses on standardization have been introduced in more than 200 educational institutions (Hesser, 2014).

As one example China Jiliang (Metrology) University (CJLU) is a university dedicated to quality supervision, inspection and quarantine. In 2017, it had 15,116 undergraduate students, 1496 graduate students, and 5800 students from independent faculties, along with 1276 full-time teachers. CJLU aims at promoting higher education for standardization by supporting publishing textbooks, exchanging programs, and researches in collaboration with partner universities' scholars, etc.[4] Another example is Tokyo Institute of Technology where courses on standardization are offered to the graduate major in Engineering Sciences and Design[5] and the Academy for Co-Creative Education of Environment and Energy Science.[6]

The Pacific Area Standards Congress (PASC) is an independent and voluntary organization of Pacific Rim National Standards Bodies. PASC has been endeavoring to provide guidelines for education on standardization. For instance, its first guidelines were prepared by Korean Standards Association and published in 2008, which

---

[4]China Jiling University, http://www.cjlu.edu.cn/.

[5]"Standardization Strategy for Global Business" (グローバルビジネスのための標準化戦略), by Graduate major in Engineering Sciences and Design (エンジニアリングデザインコース), Tokyo Institute of Technology. http://www.ocw.titech.ac.jp/index.php?module=General&action= T0300&GakubuCD=2&GakkaCD=321902&KeiCD=19&course=2&KougiCD=201708788& Nendo=2017&lang=EN&vid=03.

[6]"Global Business Strategy and Standardization & Intellectual Property" by Academy for Co-Creative Education of Environment and Energy Science, Tokyo Institute of Technology. http:// www.ocw.titech.ac.jp/index.php?module=General&action=T0300&GakubuCD=00&GakkaCD= 400041&KeiCD=0&course=41&KougiCD=201705317&Nendo=2017&lang=EN&vid=03.

featured "Case Studies of How to plan and Implement Standards Education Programs and Strategic Curriculum Model". Afterwards, one trail program that entailed 14 institutions in six countries (China, Indonesia, Japan, Korea, Peru and Vietnam) was jointly funded by Korean Agency for Technology and Standards and APEC; the result of this program was presented in "APEC Education Guideline 4: Teaching Standardization in Universities: Lessons Learned from Trial Program".[7]

4.2 Europe

In Europe, according to the above-mentioned survey by the Helmut Schmidt University, the emphasis of the courses related to standards varied considerably from university to university, and ranged from standardization governance, strategic aspects of standardization to development of IT standards and e-business applications. In 2008, the estimated number of students in Western Europe who had received these courses was approximately 500. And the situation in that year of education about standards was characterized as: very limited academic infrastructure in the field of standardization, no common curriculum, and no recognized textbooks.

A follow-up survey carried out in 2013, which was based on a limited number replies from European countries, confirmed the above-mentioned tendencies with certain positive exceptions: in Bulgaria, 9 courses were taught involving 285 students; the Rotterdam School of Management, Erasmus University (Rotterdam) offered 6 courses attended by 134 students. And an interesting and successful approach was used in Denmark where individual lectures with a focus on standards-related issues (including patents, social responsibility, IT security) were introduced in higher level education programmes. In 2010, in Denmark there was a total of 86 courses for a total of 1290 students, which increased to 177 courses and 1760 students in 2012 (Hesser, 2014).

If the results of this analysis are restricted only to specialized standardization programmes, then the total number of students trained in this area in Europe in the mid-2010s could be estimated at around 1000 per year (Hesser, 2014).

Information collected by the UNECE secretariat shows that universities are providing specialized master programmes and/or training courses. The following are a few examples: University of Geneva, Switzerland[8]; Coventry University, UK[9]; Technical University of Berlin, Germany[10]; University of Technology of Compiègne,

---

[7]See "Standards Education", PASC. https://pascnet.org/standards-education/.

[8]See "Master in Standardization, Social Regulation and Sustainable Development", University of Geneva. http://www.standardization.unige.ch/.

[9]A foundation degree programme as well as various courses. See "Metrology courses", Coventry University. http://www.coventry.ac.uk/research/areas-of-research/manufacturing-materials-engineering/metrology/metrology-courses/.

[10]"Strategic Standardization and Platform Management", Technical University of Berlin. http://www.inno.tu-berlin.de/menue/teaching/summer_term_2017/strategic_standardization/.

UNECE Initiatives on Education on Standardization   147

France[11]; University of Belgrade, Serbia[12]; and Delft University of Technology, Netherlands.[13]

Training that ranges from e-learning, themed courses to certified vocational training are also provided by national standardization bodies, such as Association Française de Normalisation (France),[14] Ente Nazionale Italiano di Unificazione (Italy),[15] Bureau de Normalisation (Belgium)[16] and Netherlands Standardization Institute, to name a few.[17]

4.3 Russian Federation and Russian-Speaking Countries

Countries of Central, Eastern and South-Eastern Europe have comparatively better knowledge on issues related to standardization than other regions at similar level of development. This situation is probably due to the mandatory character of standards during the Soviet period in the former socialist economies. Standardization was at that time included in curricula in the Soviet Union, through a special state-backed system of education on standards (under the state agency responsible for setting and implementing standards). After dissolution of the Soviet Union these educational institutions which existed practically in all ex-USSR states continued to provide such training.

There is no reliable data on the overall number of standards-related courses and/or the overall number of students who have received such training in Russian-speaking countries, however there is data on specific schools, programmes and universities.

The largest of such schools is in Moscow—Academy of Standardization, Certification and Metrology (ASCM); it has 12 branches throughout Russia and provides various post-graduate courses (ranging from a week for top managers to 2 years for experts on metrology). The total number of experts who received various forms of training at ASCM until 2015 was around 1.5–2 thousand every year (Pankina, 2015). ASCM specializes in teaching standardization as additional education with a rele-

---

[11] Various courses for master program "Quality Manager: From Strategy to Operations". See "Manager par la qualité: de la stratégie aux opérations", University of Technology of Compiègne. https://www.utc.fr/formation-continue-et-vae/performance-des-organisations/qualite.html#-tab2.

[12] Various courses for Master in Technical Legislation and Quality Management. See "Technical Legislation and Quality Management" by Faculty of Mechanical Engineering, Technical University, Sofia. http://oldweb.tu-sofia.bg/eng_new/ECTS/ectas/ectas10-11/Tabl-ECTAS-en.html.

[13] A simulation exercise "Setting Standards" by Delft University of Technology and United Knowledge. http://www.setting-standards.com/.

[14] "Training", Association Française de Normalisation. https://www.afnor.org/en/training/.

[15] "Training Centre" (Centro di Formazione), Ente Nazionale Italiano di Unificazione. http://www.uni.com/index.php?option=com_content&view=article&id=355&Itemid=2660.

[16] "NBN Academy", Bureau for Standardisation (Bureau de Normalisation). https://www.nbn.be/en/understanding-standards/nbn-academy.

[17] "Training" (Trainingen), Netherlands Standardization Institute. https://www.nen.nl/Trainingen.htm.

vant diploma. Similar training exists, for example, in the Belarus-Belarusian State Institute for Qualification Improvement and Retraining of Staff on Standardization, Metrology and Quality Management, Minsk.[18]

Specialized programmes are relatively common among Russian speaking countries as well. A few examples of schools with specialized standards-related programs are: The People's Friendship University of Russia (Russia),[19] the Belarusian National Technical University, (Belarus),[20] the Lviv Polytechnic National University (Ukraine),[21] and the Satbayev Kazakh National Technical University(Kazakhstan),[22] among others.

Some educational institutions (primarily in engineering areas) in Russia and the Commonwealth of Independent States teach separate standardization courses; such as the Tomsk Polytechnic University in Russia.[23]

4.4 North America

Also in North America, there is no reliable data regarding the overall situation on education about standards. A study in 2010 showed that in the Unites States (US), "most standards education programs have been part of on-the-job training programs at corporations, standards development organizations, private and public sector groups"(Purcell, 2010). The US education institutions that provide relevant training include the School of Engineering and Applied Science of the George Washington University,[24] the College of Applied Science and Technology of the Rochester

---

[18]Belarusian State Institute for Qualification Improvement and Retraining of Staff on Standardization, Metrology and Quality Management (Белорусский государственный институт повышения квалификации и переподготовки кадров по стандартизации, метрологии и управлению качеством), http://bgipk.by/.

[19]"Bachelor degree in Standardization and Metrology" (Стандартизация и метрология), People's Friendship University of Russia (Российский университет дружбы народов). http://www.rudn.ru/ab/bac_stand.

[20]Various degree from programmes from bachelor to doctoral by "Department of Standardization, Metrology and Information Systems" (Стандартизация, метрология и информационные системы), Belarusian National Technical University (Белорусский национальный технический университет). http://www.bntu.by/psf-smis/item/psf-smis.html.

[21]"Bachelor in Metrology, Standardization and Certification", Lviv Polytechnic National University. http://edu.lp.edu.ua/en/napryamy/6051002-metrology-standardization-and-certification.

[22]"Master's programs—Standardization and certification", Satbayev Kazakh National Technical University. http://kaznitu.kz/en/admission/gr/specialities/standardization-certificationon.

[23]"Metrology, Standardization and Certification", Tomsk Polytechnic University. http://iie.tpu.ru/en/disc_AE/mechanical_engineering/Metrology_Standardization_and_Certification.php.

[24]"Global Connections: Standards in Technology, Business and Public Policy" offered by Master of Science, George Washington University. https://eem.seas.gwu.edu/master-science.

Institute of Technology,[25] the College of Engineering and Applied Science of the University of Colorado[26] and the School of Engineering of the Catholic University of America.[27]

The National Institute of Standards and Technology (NIST), a non-regulatory agency of the US Department of Commerce also has a large range of activities aimed at promoting standards-related education. NIST offers free training to "staff of federal, state, and local agencies who need to navigate or participate in standardization".[28] Another program, the "NIST Standards Services Curricula Development Cooperative Agreement Program" provides financial assistance and support to higher education institutions which integrate standards-related content into their curricula.[29]

In Canada, the Standards in Education program (SiEp) offered by the Standards Council of Canada (SCC) enables educational institutions to have access to the ISO and IEC standards database. Over 30 educational institutions across Canada, including the University of Toronto and the University of British Columbia, are already involved in SCC's SiEp.[30]

4.5 Efforts by Regional and International Organizations

In the European Region, the three European Standardization Organizations[31] joined forces in 2012 and adopted a Masterplan on Education about Standardization, which proposed three work streams of the approach to education about standardization: building capacity, engaging key stakeholder and reaching target groups.[32]

In 2013, a Joint Working Group on Education about Standardization was formed. The working group developed a model curriculum for higher education and voca-

---

[25]"Mechanical and Electrical Controls and Standards" offered by Master in Environmental, Health and Safety Management, Rochester Institute of Technology. https://www.rit.edu/cast/cetems/ms-environmental-health-safety-management.

[26]"Advanced Topics in Quality Systems/Engineering", University of Colorado. https://www.colorado.edu/graduateschool/distance-education/course-offerings/comprehensive-course-list/emen-5043-advanced-topics-quality.

[27]Course "Strategic Standardization", by School of Engineering, Catholic University of America. http://mechanical.cua.edu/courses/descriptions-engr.cfm.

[28]See "Federal Agency Training", NIST. Available at: https://www.nist.gov/standardsgov/what-we-do/education-training/federal-agency-training.

[29]See "NIST Standards Services Curricula Development Cooperative Agreement Program", NIST. Available at: https://www.nist.gov/standardsgov/what-we-do/education-training/curricula-grants.

[30]See "Standards in Education," Standards Council of Canada. Available at: https://www.scc.ca/en/standards/obtain-standards/standards-in-education.

[31]The European Committee for Standardization (CEN), the European Committee for Electrotechnical Standardization (CENELEC) and the European Telecommunications Standards Institute (ETSI).

[32]"Masterplan on Education about Standardization", CEN, CENELEC, and ETSI. Available at: https://www.cencenelec.eu/standards/Education/JointWorkingGroup/Documents/Masterplan%20on%20Education%20about%20Standardization.pdf .

tional training respectively, and compiled a repository of tools and materials. CEN, CENELEC and ETSI also jointly held conferences on education about standards in 2012 and 2013; these conferences concluded that there was a need to devote increasing resources to education on standardization to sensitize businesses and school graduates, and a need for more coordination,[33] ignoring the earlier decision to join forces at the European level to stimulate and support the national level. Noting that Education about Standardization in Europe falls mainly under national responsibilities and competences the Joint Working Group was disbanded in 2016.

In addition to the three European Standardization Organizations, academia and researchers, are also contributing to promoting education on standards. The European Academy for Standardization (EURAS) is an example. It provides a platform for multidisciplinary exchange among standardization researchers from various academic fields and aims at promoting research and education about standardization.[34] The Belt and Road University Alliance on Standardization is a Chinese initiative to stimulate education and research in the field of standardization.

At the international level, the platform of the International Cooperation for Education about Standardization (ICES) organizes yearly workshops and other programs which gather industry, academia and standards organizations, so that "we cooperate to promote education about standardization and manage its realization".[35] Its annual meeting in 2017 is an illustration. The meeting on "Standards in Context: Skill Requirements, Pedagogy and Research" was opened by "Defining standards skill requirements: government and industry perspective"—presentations by companies and national authorities, followed by discussions by various educational institutions from Asia, Europe and America.[36]

5 Work Carried Out by UNECE on Educational Programmes on Standardization

5.1 UNECE Recommendation on Education on Standards and "StaRT-ED"

In view of the interest in teaching standardization issues of Member States and stakeholders, in November 2012, the UNECE WP.6 held an international workshop on "introducing standards-related issues in educational curricula", with approximately 100 delegates from governmental authorities, standards bodies, international orga-

---

[33] See "Joint Working Group on Education about standardization." CEN and CENELEC, https://www.cencenelec.eu/standards/Education/JointWorkingGroup/Pages/default.aspx.

[34] See "What is EURAS?" EURAS. http://www.euras.org/web1/.

[35] "About ICES", ICES. http://www.standards-education.org/.

[36] See "ISEC 2017 (Chicago)", ICES. http://www.standards-education.org/workshops/ices-2017-chicago.

nizations, the academic community, and business, from more than 20 countries. Participants noted the growing demand for specialists with knowledge and skills of standards, and at the same time regretted that very few programmes on standardization matters existed in the UNECE region.

As the result of the discussions, the WP.6 adopted a revised "Recommendation I" (on education on standardization) stressing the importance of promoting and teaching standardization through elaborating coordinated approaches and building on existing national best practices.[37]

To monitor developments in the education area within the existing UNECE team of experts—Standardization and Regulatory Techniques ("STaRT"), in 2012 a subgroup—"STaRT-ED"—representing academia was established. During the consecutive years the STaRT-ED Team organized experience exchange among universities on teaching standards. A practical tool for introducing standardization into current curricula, the "UNECE model programme on education on standardization" was endorsed, which will be discussed in the next section.

5.2 Model Programme on Education on Standardization

In 2011–2012, the UNECE secretariat researched how standardization issues were being taught in various countries in bachelor and master programmes at non-technical faculties. Besides quantitative data on the number of students learning standards in various countries, the UNECE secretariat also compared existing educational programmes. The analysis showed that programmes differed significantly in terms of content and the duration, which made their comparison difficult. If we consider the content, for example, "in Western Europe standardization is dealt with primarily from the point of view of companies, whereas in countries of Eastern Europe, Caucasus and Central Asia it is dealt with from the point of view of regulatory authorities. Almost none of the existing programmes deal with flanking issues such as metrology and market surveillance"(UNECE, 2012a, b)."[page 2]

To address the diverging educational angles and approaches, the UNECE secretariat, in consultation with academia, prepared a proposal for a "model programme on education on standardization"[38] with a common minimum set of standardization topics that a graduate should master from a business and governmental perspective. That is, this set includes the major standardization, regulatory and related issues relevant to the activities of business and of regulatory and administrative authorities.

---

[37] UNECE document ECE/TRADE/C/WP.6/2012/6, Concept note for the Workshop on "Introducing standards related issues in educational curricula", including a proposed model programme on standardization" available at https://www.unece.org/fileadmin/DAM/trade/wp6/documents/2012/wp6_2012_06E.pdf.

[38] UNECE document ECE/TRADE/C/WP.6/2012/6. "Concept note for the Workshop on "Introducing standards related issues in educational curricula", including a proposed model programme on standardization", UNECE, ECE/TRADE/C/WP.6/2012/6, 2012. Available at: http://www.unece.org/fileadmin/DAM/trade/wp6/documents/2012/wp6_2012_06E.pdf.

The programme targets general academic programmes at university level rather than that it provides specialized in-depth training on standardization. The agreed essential issues are grouped into 15 modules. The programme can be used either in full as a stand-alone full-fledged course; or in separate modules for training on specific subjects. The teaching time recommended for modules may vary depending on the purpose of training and on the entry level of knowledge of students.

One of the leading global educational institution in the area of standards, has prepared, based on the UNECE model programme, the first three teaching modules which were presented to the STaRT-ED Team and were endorsed by the WP.6 at its November 2014 annual session.

These modules (on the benefits of standardization; standards in the WTO context and on quality management systems) are freely available to any university interested in teaching standards, and they mark the de facto launching of the UNECE educational programme as a practical teaching tool.[39]

Because the UNECE model programme is module-based, it provides flexibility for professors while at the same time ensuring that a graduate gains at least the minimum knowledge of the subject. For example, module 14 is devoted to a practical exercise to show how standards can be incorporated into a company's strategy and how they can support procurement, production, etc. These issues are important, for example, for graduates who will work in a manufacturing plant. This module is suggested for inclusion when teaching students without prior knowledge of such issues in general education curricula. Thus, this module might be optional or mandatory depending on the entry level of relevant student groups.

Currently, the transfer of credits between programmes or institutions is primarily based on the number of instruction hours instead of on its content or learning outcome. The UNECE model programme not only provides educational institutions with a check-list of essential standards-related issues, but also marks a first step towards the harmonization of educational programmes and eventually towards the mutual recognition of acquired skills, curricula and diplomas.

One of the practical results of these awareness raising and capacity building efforts by UNECE was the introduction of new courses on standards in certain regions and wider attention to standards issues within existing programmes.[40] For example, the Moscow State Regional University (MSRU) introduced in 2013 the new course "fundamentals of standardization"; its content (15 themes) follows the thematic areas of the UNECE model programme.[41] 273 students and 254 students received supplementary training on standardization matters in 2013–2014 and 2015–2015 respectively. Moreover, according to MSRU, students evaluated the new course as useful (UNECE,

---

[39]See "Thematic Areas—Education", UNECE, https://www.unece.org/tradewelcome/steering-committee-on-trade-capacity-and-standards/tradewp6/thematic-areas/education.html.

[40]Information from the UNECE Secretariat.

[41]Moscow State Regional University, https://mgou.ru/en/.

2014).[42] Another positive example is Matej Bel University, Slovakia, where a course "standardization and international standards is offered as part of the master degree programme on "finance, banking and investment".[43] The content of this course also roughly follows the UNECE model programme. It consists of 13 areas, only two of which differ from UNECE's model.

The model program will facilitate the inclusion of introductory and awareness building courses on standards in institutions which may currently lack qualified expertise on this subject. Interest was also expressed in the development of similar teaching modules on standards-related issues in other areas, such as for example transport and customs.

Finally, the UNECE Secretariat has also a dedicated webpage that contains a library of material for teaching standards from various universities. These are available to educational institutions interested in launching a similar training. The website also contains links to the work done by other organizations, such as ISO (repository of teaching materials and studies on the benefits of standardization) and APEC (on teaching standardization in universities), among others. [44]

In order to help universities introduce standardization courses, UNECE in cooperation with academia initiated a project on the preparation of teaching materials. [confusing; how does the model (Sect. 5.2) relate to this one?].

6 Conclusions and Directions for Future Work

Standards are playing a growing role in the international arena, as a trade facilitation and a trade distortion factor, as well as an enabler of sustainable development. Standards and related regulatory tools are increasingly influencing the competitive position of companies and countries. Knowledge of standards is an important part of the portfolio of competencies and skills that have been and will continue to be required for job seekers in engineering professions. Recent studies show that large companies are starting to understand that awareness of standards are even an asset for non-engineer employees dealing with the marketing of products.

In this context, the UNECE recommendation on introducing standards-related issues in educational curricula is well-advised, and should be further promoted. Awareness of the importance of standards is beneficial to the society as a whole as well as to graduates seeking jobs. Education on standards is useful not only for higher educational institutions, which UNECE activities focus on, but also for primary

---

[42]See "Results on introducing standardization discipline, MSRU." UNECE. Available at: http://www.unece.org/fileadmin/DAM/trade/wp6/AreasOfWork/EducationOnStandardization/MGOU-introduction_of_standardization_discipline_results_Nov2014.pdf.

[43]By Faculty of Economics, Matej Bel University, http://www.ef.umb.sk/index_e.asp.

[44]See "Thematic Areas—Education", UNECE, https://www.unece.org/tradewelcome/steering-committee-on-trade-capacity-and-standards/tradewp6/thematic-areas/education.html.

and secondary schools. The education content and pedagogic methods should be adequately tailored to the specific needs of each target group.

Current state education on standardization, particularly in Europe, is unsatisfactory and unable to match market and regulatory demands. There are various reasons for the weak response of educational institutions towards introducing standardization into curricula, which include: often ambiguous requirements from business (i.e., standardization needs are not clearly articulated); lack of nationally, regionally, and internationally agreed programme content; lack of trained professors and of agreed textbooks; and specific national conditions relating to educational systems, e.g. the organization of the national education system.

The approach to teaching standardization also needs to be reviewed. Traditionally standardization was part of curricula of engineering professions or as a part of vocational training, and was usually highly specialized and linked to a specific professional profile. The appearance of new standards that go beyond technical matters such as ecology, social, etc. shows the necessity to recognize and understand the multidisciplinary character of standardization and to teach it accordingly.

Relevant directions for future work can be summarized as follows:

(1) **Continue awareness building.** The fact that the education on standardization is insufficient to meet market and regulatory demands shows that continuing awareness building is indispensable; much more needs to be done. Relevant organizations and authorities can help industries and governments to communicate their needs more clearly to the public and to educational institutions.

(2) **Involve multiple stakeholders.** The major stakeholders in the area of education on standards include: companies, employees, ministries of education, universities, students, regulatory authorities and policy makers, and standards-setting bodies (national, regional, and international). In the future UNECE is planning to involve all these groups in the process of identifying and building a portfolio of desired competencies for graduating students, company employees and public administration officials. It is in the process of preparing training programmes and materials.

(3) **Support synergies and further cooperation.** The above analysis shows that many organizations, institutions, and other groups of actors have been engaging in parallel and similar activities related to standards. The challenge is how to orchestrate such activities and efforts.

(4) **Build capacity**. UNECE and other organizations can help relevant institutions build up expertise for teaching standards at educational institutions, and elaborate the training materials of teaching programmes.

(5) **Tailor messages to different audiences.** For example, education on standardization is significantly influenced by the way national education systems are organized. In countries with a relatively centralized approach, where education ministries influence the content of the universities' educational programmes (i.e. Russia), efforts should start at the level of ministries. In countries where universities independently decide on education content (i.e. UK), it is necessary to approach each university individually and to persuade it of the importance

and usefulness of adding this area to curricula (which is evidently more complicated and time consuming.) The UNECE is prepared to and has been working in both such directions.

(6) **Look into new standards-related areas**. Other than the traditional areas, i.e. engineering, that have been associated with standards, standards are for instance becoming increasingly important for sustainability and the environment. This is recognized by the theme of the 27th annual session of the UNECE WP.6, which features an International Conference on "Standards for the Sustainable Development Goals" and the cross-sectoral discussions that have brought these new standards-related areas in the spotlight.

Annex I

UNECE Recommendation I
Education on standards-related issues[45]

The Working Party on Regulatory Cooperation and Standardization Policies,

Recognizing the role and place of standards and of quality infrastructure in accompanying or controlling products during their life cycle,

Underlining the important contribution of standards and regulatory framework (technical regulations, metrology, conformity assessment, accreditation, market surveillance) in the attainment of national and international development goals (including the United Nations Millennium Development Goals) and in promoting sustainable development,

Recommends that—in collaboration with appropriate intergovernmental and other organizations and academia, and taking into account the activities of global, regional and national standards bodies—Governments should encourage, wherever feasible and where the national legal framework permits:

(a) the introduction by educational establishments of the subject of standardization into the curricula of educational establishments and particularly of universities for students majoring in technical and scientific subjects, as well as in legal, economic and management studies;
(b) the vocational education and training of specialists in standardization;
(c) the enhancement of awareness-raising activities targeted to the business community and regulatory authorities (in particular, trade and customs officials);
(d) the further study of standardization issues in order to identify best practices in ensuring that standardization and regulatory regimes contribute to meeting the legitimate concerns of society (e.g. human safety, environment) without creating unnecessary technical barriers to trade.

---

[45]Working Party on Regulatory Cooperation and Standardization Policies. Recommendation adopted in 1970 and revised in 2012. Available at: http://www.unece.org/fileadmin/DAM/trade/wp6/Recommendations/Rec_I.pdf.

Annex II

The UNECE Proposed model programme on standardization includes the following modules[46]:

1. Standardization basics
2. Benefits of standardization for society
3. Standardization and companies
4. National legal and institutional framework
5. Regulatory policies and related institutional mechanisms
6. Managing risks through standards, regulations and regulatory impact assessments (RIAs)
7. Metrology
8. Conformity assessment
9. Market surveillance
10. Management system standards
11. International standardization
12. International trade, standards and regulations
13. Standardization of information requirements and supply chains
14. A practical exercise: Standardization within a company
15. Policy issues and challenges in standardization.

Bibliography

Association Française de Normalisation. (n.d.). Training. Retrieved November 2, 2017 from https://www.afnor.org/en/training/.

Белорусский государственный институт повышения квалификации и переподготовки кадров по стандартизации, метрологии и управлению качеством [Belarusian State Institute for Qualification Improvement and Retraining of Staff on Standardization, Metrology and Quality Management]. (n.d.). [Home page]. Retrieved November 2, 2017 from http://bgipk.by/.

Белорусский национальный технический университет [Belarusian National Technical University]. (n.d.). Кафедра « Стандартизация, метрология и информационные системы » [Department of Standardization, Metrology and Information Systems]. Retrieved November 2, 2017 from http://www.bntu.by/psf-smis/item/psf-smis.html.

Bureau de Normalisation [Bureau for Standardisation]. (n.d.). NBN Academy. Retrieved November 2, 2017 from https://www.nbn.be/en/understanding-standards/nbn-academy.

Catholic University of America. (n.d.). Strategic Standardization by Mechanical Engineering. Retrieved November 2, 2017 from http://mechanical.cua.edu/courses/descriptions-engr.cfm.

CEN and CENELEC. (n.d.). Joint Working Group on Education about standardization. Retrieved November 2, 2017 from https://www.cencenelec.eu/standards/Education/JointWorkingGroup/Pages/default.aspx.

CEN, CENELEC, & ETSI. (2012). Masterplan on Education about Standardization. Retrieved from https://www.cencenelec.eu/standards/Education/JointWorkingGroup/Documents/Masterplan%20on%20Education%20about%20Standardization.pdf.

---

[46]For details see "Proposed model programme on standardization", supra note 39.

UNECE Initiatives on Education on Standardization

CEN. (n.d.). [Home page]. Retrieved November 2, 2017 from https://www.cen.eu/Pages/default. aspx.

CENELEC. (n.d.). [Home page]. Retrieved November 2, 2017 from https://www.cenelec.eu/.

China Jiling University. (n.d.). [Home page]. Retrieved November 2, 2017 from http://www.cjlu. edu.cn/.

Coventry University. (n.d.). Metrology Courses. Retrieved November 2, 2017 from https://www. coventry.ac.uk/research/areas-of-research/manufacturing-materials-engineering/metrology/ metrology-courses/.

Delft University of Technology and United Knowledge. (n.d.). Setting Standards. Retrieved November 2, 2017 from http://www.setting-standards.com/.

Ente Nazionale Italiano di Unificazione [Italian Organization for Standardization]. (n.d.). Centro di Formazione [Training Centre]. Retrieved November 2, 2017 from http://www.uni.com/index. php?option=com_content&view=article&id=355&Itemid=2660.

ETSI. (n.d.). [Home page]. Retrieved November 2, 2017 from http://www.etsi.org/.

EURAS. (n.d.). What is EURAS. Retrieved November 2, 2017 from http://www.euras.org/web1/.

European Union law. (2012, October 25). EU resolution No. 1025/2012. Retrieved from http://eur-lex.europa.eu/LexUriServ/LexUriServ.do?uri=OJ:L:2012:316:0012:0033:EN:PDF.

George Washington University. (n.d.). Global Connections: Standards in Technology, Business and Public Policy by Master of Science. Retrieved November 2, 2017 from https://eem.seas.gwu.edu/ master-science.

Hesser, W. (2014). Memorandum on Standardization in higher Education in Europe. Retrieved from http://www.iso.org/sites/edumaterials/hesser-memorandum.pdf.

ICES. (n.d.). About ICES. Retrieved November 2, 2017 from http://www.standards-education.org/ about-ices.

ICES. (n.d.). ICES 2017 (Chicago). Retrieved November 2, 2017 from http://www.standards-education.org/workshops/ices-2017-chicago.

IFAN. (2014). Education and Training about Standardization. Retrieved from http://www.ifan.org/ IFAN-GUIDE%204-Education-2014-09.pdf.

Lviv Polytechnic National University. (n.d.). Bachelor in Metrology, Standardization and Certification. Retrieved November 2, 2017 from http://edu.lp.edu.ua/en/napryamy/6051002-metrology-standardization-and-certification.

Matej Bel University. (n.d.). Faculty of Economics. Retrieved November 2, 2017 from http://www. ef.umb.sk/index_e.asp.

Moscow Region State University. (n.d.). [Home page]. Retrieved November 2, 2017 from https:// mgou.ru/en/.

NEN. (n.d.). Trainingen [Training]. Retrieved November 2, 2017 from https://www.nen.nl/ Trainingen.htm.

NIST. (2016, May 16). Federal Agency Training. Retrieved November 2, 2017 from https://www. nist.gov/standardsgov/what-we-do/education-training/federal-agency-training.

NIST. (2016, May 16). NIST Standards Services Curricula Development Cooperative Agreement Program. Retrieved November 2, 2017 from https://www.nist.gov/standardsgov/what-we-do/ education-training/curricula-grants.

Pankina, G. (2015, November). Presented at the START-ED meeting, Geneva.

PASC. (n.d.). Standards Education. Retrieved November 2, 2017 from https://pascnet.org/standards-education/.

Российский университет дружбы народов [People's Friendship University of Russia]. (n.d.). Стандартизация и метрология [Bachelor degree in Standardization and Metrology]. Retrieved November 2, 2017 from http://www.rudn.ru/ab/bac_stand.

Purcell, D. (2010). The Strategic Value of Standards Education. *ISO*. Retrieved from http://www. iso.org/sites/materials/initiatives-in-education/education_initiatives-higher-edu/educational_ materials-detail-initiatives4dfe.html?emid=36.

Rochester Institute of Technology. (n.d.). Mechanical and Electrical Controls and Standards by MS in Environmental, Health & Safety Management. Retrieved November 2, 2017 from https://www.rit.edu/cast/cetems/ms-environmental-health-safety-management.

Satbayev Kazakh National Technical University. (n.d.). Master's programs—Standardization and certification. Retrieved November 2, 2017 from http://kaznitu.kz/en/admission/gr/specialities/standardization-certificationon.

Standards Council of Canada. (2012, March 7). Standards in Education. Retrieved November 2, 2017 from https://www.scc.ca/en/standards/obtain-standards/standards-in-education.

Technical University of Berlin. (n.d.). Strategic Standardization. Retrieved November 2, 2017 from http://www.inno.tu-berlin.de/menue/teaching/summer_term_2017/strategic_standardization/.

Technical University, Sofia. (n.d.). Technical Legislation and Quality Management by Faculty of Mechanical Engineering. Retrieved November 2, 2017 from http://oldweb.tu-sofia.bg/eng_new/ECTS/ectas/ectas10-11/Tabl-ECTAS-en.html.

Tokyo Institute of Technology. (n.d.). Global Business Strategy and Standardization & Intellectual Property. Retrieved November 2, 2017 from http://www.ocw.titech.ac.jp/index.php?module=General&action=T0300&GakubuCD=00&GakkaCD=400041&KeiCD=0&course=41&KougiCD=201705317&Nendo=2017&lang=EN&vid=03.

Tokyo Institute of Technology. (n.d.). Standardization Strategy for Global Business. Retrieved November 2, 2017 from http://www.ocw.titech.ac.jp/index.php?module=General&action=T0300&GakubuCD=2&GakkaCD=321902&KeiCD=19&course=2&KougiCD=201708788&Nendo=2017&lang=EN&vid=03.

Tomsk Polytechnic University. (n.d.). Metrology, Standardization and Certification. Retrieved November 2, 2017 from http://iie.tpu.ru/en/disc_AE/mechanical_engineering/Metrology_Standardization_and_Certification.php.

UNECE (2012). Document ECE/TRADE/C/WP.6/2012/7: Draft revised recommendation "Education on standards-related issues. Retrieved from https://www.itu.int/en/ITU-T/academia/Documents/stdsedu/1st%20Meeting-20121008-Denmark/004_Att.1-AHG_SE_UNECE_Rec.StdsEdu.pdf.

UNECE, and OECD. (2016). International Regulatory Cooperation: The Case of the United Nations Economic Commission for Europe. Retrieved from https://www.unece.org/index.php?id=44268.

UNECE, ISO, IEC, and ITU. (2000, March). Memorandum of Understanding on electronic business. Retrieved November 2, 2017 from http://www.itu.int/ITU-T/e-business/files/mou.pdf.

UNECE. (2012). Document ECE/TRADE/C/WP.6/2012/6: Concept note for the Workshop on "Introducing standards related issues in educational curricula", including a proposed model programme on standardization. Retrieved from http://www.unece.org/fileadmin/DAM/trade/wp6/documents/2012/wp6_2012_06E.pdf.

UNECE. (2014, November). Results on introducing standardization discipline, MSRU | UNECE. Retrieved November 2, 2017 from http://www.unece.org/fileadmin/DAM/trade/wp6/AreasOfWork/EducationOnStandardization/MGOU-introduction_of_standardization_discipline_results_Nov2014.pdf.

UNECE. (n.d.). [Home page]. Retrieved November 3, 2017 from https://www.unece.org/info/ece-homepage.html.

UNECE. (n.d.). Thematic Areas—Education. Retrieved November 2, 2017 from https://www.unece.org/tradewelcome/steering-committee-on-trade-capacity-and-standards/tradewp6/thematic-areas/education.html.

University of Bremen, and DFKI–German Research Centre for Artificial Intelligence. (2012). Standards Engineers. Retrieved from https://www.unece.org/fileadmin/DAM/trade/wp6/AreasOfWork/RiskManagement/_notes/Survey_on_Standards_in_job_descriptions.pdf.

University of Colorado. (2017, May 25). Advanced Topics in Quality Systems/Engineering. Retrieved November 2, 2017 from https://www.colorado.edu/graduateschool/distance-education/course-offerings/comprehensive-course-list/emen-5043-advanced-topics-quality.

University of Geneva. (n.d.). Master in Standardization, Social Regulation and Sustainable Development. Retrieved November 2, 2017 from http://www.standardization.unige.ch/.

University of Technology of Compiègne. (n.d.). Manager par la qualité : de la stratégie aux opérations [Quality Manager: From Strategy to Operations]. Retrieved November 2, 2017 from https://www.utc.fr/formation-continue-et-vae/performance-des-organisations/qualite.html#-tab2.

WTO. (2012). *World Trade Report 2012, Trade and public policies: A closer look at non-tariff measures in the 21st century.* Retrieved from https://www.wto.org/english/res_e/publications_e/wtr12_e.htm.

WTO. (n.d.). *Statistics Database.* Retrieved November 2, 2017 from http://stat.wto.org/Home/WSDBHome.aspx.

Lorenza Jachia is a trade and development economist, Lorenza Jachia holds a Masters Degree from the Graduate Institute for International Studies (Geneva) and a Bachelors degree from Bocconi University (Milan). She has been working at the United Nations since 1995, at first in the United Nations Conference on Trade and Development (UNCTAD) and then in the United Nations Economic Commission for Europe (UNECE). Early in her career she provided training and advisory services to negotiators of free trade area agreements and published on the topic of regional integration. Her work then focused on the deep aspects of economic integration, in particular: overcoming technical barriers to trade through the approximation of technical regulations. Currently, the thrust of hew work is on how standards help the implementation of the 2030 agenda for Sustainable Development. Since April 2008, as the Secretary of the UNECE Working Party on "Regulatory Cooperation and Standardization Policies", she works to bring the United Nations and the "standards community" together. Lorenza is the author of over 20 publications including "Risk Management in Regulatory Systems". She has contributed to over one hundred UN reports including many interagency publications.

Serguei Kouzmine has been working for the United Nations for almost 30 years including for UNECE (UN Economic Commission for Europe) where he was engaged in standardization policies and trade and transport facilitation programmes. He is one of co-founders of the UNECE project "education on standardization" to which he has contributed extensively. His work on promoting and teaching standards has been praised by governments, educational institutions and business. Dr. Kouzmine is recognized internationally as an expert on trade and standards related issues and was the recipient of a number of national, regional, sectoral honorable awards in the field of standardization. He is an author of more than 80 various publications/articles on trade, entrepreneurship, environment and standards matters. He holds a PHD degree in economics from the Russian Market Research Institute and a master degree in international trade from Moscow University of International Relations. Dr. Kouzmine can be contacted at: serge.kouzmin@gmail.com.

Haiying Xu who was interning at the UNECE at the time of writing, provided research assistance to this paper. She obtained her Master Degree from the Graduate Institute for International Studies (Geneva) and her Bachelor Degree in Laws from Renmin University of China (Beijing). Prior to her internship at the UNECE, she has also interned at the International Centre for Trade and Sustainable Development (ICTSD) and the International Trade Centre (ITC, joint agency of the UN and the WTO).

The Need for Multi-disciplinary Education About Standardization

Olia Kanevskaia

1 Introduction

We live in the world of a hyper-connected society, where human interactions are governed by a bewildering variety of rules. Harmonized technical requirements provided by global standards bodies enable economic exchange of goods and services, technical specifications and protocols issued by private actors allow data flow through electronic devices, and all this exists within the web of national and global legal frameworks. The challenges of past centuries seem to fade into the background or are replaced by concerns related to digitalization and technological advancement.

And yet, traditional cooperation barriers have not vanished completely, especially in cases where technical specifications are collectively defined by various types of stakeholders. States may still adopt restrictive measures targeting foreign imports and undermining regulatory coherence between trade partners (see one of the recent examples in WTO, 2017). Companies may opt to define widely applicable technical specifications in consortia with limited membership, rather than formalized global standards bodies with established reputation, as it is the case for standards in information and communication technologies (ICT) (Delcamp & Leiponen, 2014). National suppliers may prefer to use different technical specifications than agreed among the industry: Rhineland oil refinery, for instance, uses DIN EN 2501-1 standards for flanges, while the global standards for flanges is ANSI/ASME B 16.5. But most importantly, languages spoken by industry experts still appear to impact pri-

---

TILEC's research program on economic governance has received funding from public authorities such as NMa, NZa, ACM or the European Commission as well as private companies such as Microsoft and Qualcomm. TILEC research is conducted in accordance with the rules set out in the Royal Dutch Academy of Sciences (KNAW) Declaration of Scientific Independence.

O. Kanevskaia (✉)
Tilburg Law and Economics Center (TILEC) and Tilburg Law School (TLS), Tilburg, The Netherlands
e-mail: o.s.kanevskaia@uvt.nl

© Springer Nature Switzerland AG 2020
S. O. Idowu et al. (eds.), *Sustainable Development*, CSR, Sustainability,
Ethics & Governance, https://doi.org/10.1007/978-3-030-28715-3_10

vate regulatory processes: either literally, when participants are faced with linguistic barriers during technical meetings, or figuratively, when experts disagree on grounds common only for their field of expertise.

Against this backdrop, the need for outreach and education about standardization has been acknowledged for over two decades (Hesser & Czaya, 1999; De Vries, 2002; Spivak & Kelly, 2003). In recent times, this need is strengthened by increased digitalization, shorter products' innovation cycles and expansion of markets, whereby standards are often used as instruments of economic policy and carry implications for different sectors. Despite numerous initiatives to promote standardization as an integral part of educational curriculum in engineering, management and business studies (i.e. Hesser & Czaya, 1999; De Vries, 2002; Spivak & Kelly, 2003; De Vries & Egyedi, 2007; Hesser & Siedersleben, 2007; Hesser, 2014), academic endeavors on *legal* education about standardization remain particularly scarce. Yet, ignoring fundamental legal concepts in standardization courses may cause misunderstandings and even deter cross-sectoral and multinational cooperation.

The story of the Tower of Babel suggests that cooperation leading to the provision of public good is impossible without stakeholders speaking in a common language. This implies that when defining a standard, technical, social, economic, legal, and even organizational concerns should be taken into account. Following this reasoning, education about standardization should offer insights of a wide range of disciplines, rather than merely focusing on a single scientific or academic field of study (Hesser & Siedersleben, 2007). Such an approach will facilitate future cooperation among standardization professionals and in the long run, will improve the quality of standards.

While explaining the necessity of a multi-disciplinary approach to standardization and education about standardization (Sect. 2) and providing examples of experts' disagreements on various standardization concepts, drawing upon actual conversations with engineers, business analysts and software developers (Sect. 3), this contribution suggests that increased multi-disciplinary education on standardization provides the recipe for smoothing out the differences between standardization sectors and assures an efficient and inclusive standards development processes (Sect. 4). This paper also proposes a number of methods that could be implemented by academic institutions, national and international standards bodies and private sector enterprises to increase awareness and understanding of cross-sectoral standardization (Sect. 4). It concludes on an optimistic note that due to the growing multi-disciplinary research on standardization, multi-diciplinary education about standardization will also be strengthened, but warns that academic curricula should be designed taking into account the overall goals of educational programs and, additionally, the students' and professionals' understanding of the use of standards in other disciplines (Sect. 5).

2 Multi- and Inter-disciplinary Standardization: A Legal Perspective

2.1 Pluralistic Communities and Private Norm-Making

The globalized society of today does not provide space for a monodisciplinary approach. Managers are required to take into account a myriad of legal issues, such as trade sanctions and anti-bribery policies, when performing their duties. In patent disputes, judges often adjudicate on what has traditionally been assumed an engineering matter. It is common for social sciences to be translated to legal setting by field experts in courtrooms and (international) tribunals. Business analysts often have to give due considerations to the regulatory environment in which businesses operate. Physicians may be required to operate high-tech equipment to provide the necessary care for their patients.

For a long time, scholars have been cherishing the idea that the global order is based on legal pluralism, where normative systems imposed by industry and private actors co-exist with States-driven regulation (i.e. Berman, 2012). Consider this: a company should adhere to the laws of the State of its incorporation, but also to the international rules imposed by the competent organizations, such the International Labor Organization (ILO) in Geneva; its activities are subject to widely acknowledged quality management and environmental norms, like ISO 9001 and ISO 14001 standards, but also to the company's code of conduct or other types of single actor-schemes. To allow its employees to carry out their tasks, an enterprise has to ensure the availability of working places, desks and electronic equipment. The terms of the lease of a property where a company's office is located is determined in a lease agreement; the office building has to comply with applicable building codes and corresponding safety requirements; internet connection should be provided by the means of a Wi-Fi router that follows technical specifications for the wireless local area network (WLAN).

A world where normative requirements are disseminated in plural ways would not hold on monodisciplinary solutions: the hybridity of industry rules, private arrangements and governmental regulations requires a comprehensive approach that recognizes a variety of scientific fields (Posner, 1987; Abbott, 1989; Berman, 2005; Larouche, 2012).

A vivid example of a synergy between multiple disciplines is the process of formulating and implementing technical standards. Being common agreements that define requirements for a product, design or process, standards enable information transfer and connect processes, things and people (De Vries, 1997; Krechmer, 2006); while carrying out a range of significant normative, cognitive and regulatory functions (Lane, 1997). Standards exist in a tremendous variety of forms, ranging from measurements specifications, requirements for quality control and financial reporting, to methods and protocols enabling the necessary interference and providing interoperability among the elements of a technological system. Not only do they serve as the "fillers" of "regulatory gaps" (Cassese, 2005), but they also facilitate the functioning of supply chain, reduce costs and unlock multiple possibilities for the use of modern technologies in a range of sectors (Curran, 2003). A laptop is running on about two hundred different standards, including web protocols, memory chips and WLAN specifications (Biddle, White, & Woods, 2010); a smart meter supports standardized interfaces for data communication, which enables its connection with consumers' home-automation system (Cervigni & Larouche, 2014). The ubiquitousness of standards is reflected in a rather simplified, but accurate classification offered by Brunsson and Jacobsson (2000): "standards for what we are, what we do and what we have".

Standards are typically praised for harmonizing technical requirements across the countries, reducing regulatory uncertainties and producing network effects. At a micro level, adherence to standards contribute to the growth and prosperity of a company: firms whose standards are well-accepted on the market are likely to experience high returns, while firms implementing less successful standards may end up being "locked out" of the market (Büthe, 2010). In the global context, standards may both facilitate market access and erect trade barriers and, given their allocation efficiency and distributional capacity, create and correct market failures (Büthe, 2010). Over the past decades, standards became particularly important to global society due to the increased reliance on technology and the growing importance of the Internet.

2.2 Standardization as a Field of Multi-disciplinary Research

Given standards' capacity to affect an infinite variety of fields and to have wider policy implications, it comes as no surprise that standardization has sparked the curiosity of different, often non-technical, stakeholders (Krechmer, 2007). In particular, this applies to the field of technical and scientific innovation, where the rapid emergence of multifaceted technologies fosters awareness amongst all innovation systems' actors regarding, for instance, societal infrastructure, multiplicity of technological approaches and degree of regulation (O'Sullivan & Brévignon-Dodin, 2012). Standardization thus does not only require complex technical considerations, but also organizational and strategic coordination amongst the multiplicity of stakeholders (Russel, 2006). With that said, research on standardization should ideally reflect the interplay between various academic domains. At the same time, since standardization has not (yet) developed its own theoretical foundations, its study should be based on the concepts and disciplines that it integrates (De Vries, 2015).

Indeed, multi- and inter-disciplinary research on standardization is gradually entering into the academic arena. International Journal of Standardization Research (IJSR) currently covers a broad variety of standardization topics, including management, intellectual property and history (i.e. IJSR 2017); peer-reviewed journals devote their special issues to the topics of standards-related governance (i.e. JEPP, 2001) or law and economics of technology standards (i.e. JCLE, 2013); recent research on standardization modes build on studies from multi-disciplinary fields (i.e. Wiegmann, De Vries, & Blind, 2017); and book volumes on standardization combine such disciplines as law, economics, management, political science and sociology (Peters, Koechlin, Förster, & Fenner Zinkernagel, 2009; Delimatsis, 2015; Contreras, 2018). In contrast, multi-and inter-disciplinarity has not yet been sufficiently introduced to academic curricula, leaving education about standardization rather monodisciplinary.

2.3 Past Agenda on Education About Standardization

Previous experiences in standardization committees demonstrated that many professionals working with standards lack sufficient education about standardization (De Vries, 2003). In turn, knowledge of standards and standardization was recognized as an asset for economists and lawyers (Spivak & Kelly, 2003), but also for managers (De Vries & Egyedi, 2007). Participants of earlier workshops on education about standardization agreed that education of professionals should be provided on a continuous basis and should be based on the learning objectives that correspond to their professional activities (De Vries & Egyedi, 2007).

Yet, academic education about standardization is equally important as education of professionals, since it also serves as a prerequisite for safeguarding the competitiveness of the economy and society (Hesser, 2014). In higher educational studies, calls for offering students a theoretical basis to facilitate their understanding of standardization have been growing for more than a decade (e.g. Spivak & Kelly, 2003; Krechmer, 2007); at the same time, education about standardization was believed to lack a systemic approach (De Vries, 2003) and, at least in Europe, to be short of investment in an academic infrastructure (Hesser, 2014). In the absence of a uniform concept of an EU standardization education system, higher educational programs face two options to include standardization into their curriculum: either introducing a separate course on standards and standardization or integrating standards-related topics in already existing academic classes (De Vries & Egyedi, 2007). In this regard, inter- and multi-disciplinary teaching methods have been identified as the essential components of higher education on standards (Hesser & Czaya, 1999; Spivak & Kelly, 2003).

Unfortunately, the significance of multi-disciplinary education is often underestimated by many standardization practitioners. In the beginning of my academic research on standard-setting processes, I was often confronted by professionals who, despite their incontestable expertise on a particular turf, expressed a certain amount of skepticism or even lacked awareness once it came to concerns associated with technical standards in domains other than their own. I owe a great debt to these experts, whose generosity, patience and eagerness to share their knowledge opened my eyes on the practical angle of standards' creation and implementation, but also revealed the necessity to promote multi-disciplinary education about standardization, which served as an underlying motivation for this contribution. In the following sections, I illustrate my position by discussing the statements of various experts, which struck me the most during the initial stages of my research and required an immediate clarification from my side.

3 Contested Topics on Standardization

3.1 Technical Standards Cannot Incorporate Essential Proprietary Solutions

Engineers and lawyers rarely agree, especially when it comes to patent matters. While certainly not all engineers may claim to fully apprehend legal consequences of their technical decisions, lawyers often come under fire for holding a "race to courtroom". Just think about the endless saga of patent wars, where technology giants, such as Apple, Google, Samsung and many others, got involved in a costly litigation on design and utility patents for smartphones; or about the so-called "patent trolls," with a famous example of *Innovatio IP* suing a large number of motels and local coffee shops for allegedly breaching its patents for Wi-Fi specifications.

In the realm of ICT and telecommunications, conflicts amongst stakeholders may arise when the functioning of a standard depends on technologies protected by (multiple) patents. Such functional utility patents may become essential for standards' implementation either due to the lack of any alternative methods ("technical essentiality"), or when other means of implementation are commercially or substantially inferior ("commercial essentiality"), and are commonly referred to as "Standards Essential Patents" (SEPs) (Baron & Pohlmann, 2015). To adopt a standard or a technical specification, device manufacturers thus need to obtain a "permission" from patent holders to use their proprietary technology, which is typically granted against a fee payment in a form of a license. Whereas high licensing fees prevent the access to the essential technologies and hence to the standards, contributions from patent-holders are discouraged by payments that are much lower than their expectations. To ensure that the interests of both technology-owners and implementers are respected in the processes of standards formulation, industry associations may require patent holders to disclose patent claims that may become essential for a standard and to require licensor and licensees to establish royalty rates based on Fair, Reasonable and Non-Discriminatory (FRAND) or royalty-free terms (Contreras, 2011).

Yet, in standardization of technical requirements for pressure equipment, proprietary solutions that are essential are somewhat rare. Mechanical engineers often raise their eyebrows when they come across the terms "standards" and "essential patents" in one sentence. Some of them would even consider patents an anathema that obstructs the openness of a standard. And while patenting of mechanical tools and methods is generally not uncommon, avoidance of any proprietary elements appears the preferred strategy of standardizers in this field. For a long time, patents have also been an unfamiliar phenomenon in sustainability standards, including such areas as food labelling or environment management. In the Committee on Technical

Barriers to Trade (TBT), a body within the World Trade Organization (WTO) dealing with technical regulations and standards, patents are discussed only occasionally (see examples in Wu, 2017).

It should be acknowledged that every discipline has its own policies for intellectual property (IP) management, tailored to the specific industry and operational setting: it is therefore not astonishing that patent-related issues typically occur in sectors that heavily rely on Research and Development (R&D) and technological innovation, such as wireless networks, internet and telecommunications. But industry settings are prone to changes, especially in the view of increased digitalization and continuous demand for interconnectivity. Standards for the Internet of Things, for instance, are not developed within a sole expertise domain, but are based on cross-sectoral cooperation. Likewise, the emergence of 3D printed alternatives for traditional materials and techniques prompts the question whether the approach to patents in the sectors other than ICT and telecommunication should be revisited. To meet market requirements, future standards may be based on an amalgam of proprietary and non-patented solutions and require a combination of different patent licensing structures: the current debate on open source and FRAND-based licenses provides a good illustration of this development (i.e. Lundell, Gamalielsson, & Katz, 2015); (European Commission, 2017). In this regard, addressing the nexus between standardization and patents in sectors other than ICT and telecommunications does not seem otiose, but rather lays the groundwork for fruitful and necessary symbiosis between different industries.

3.2 Private Voluntary Standards Have no Regulatory Impact

When sharing my thoughts on wireless and internet standards, the first question posed by a large share of industry experts is: "Why do you care?" Aside from addressing my personal interest in the wireless industry, this question can be interpreted as: "Why should lawyers be concerned with voluntary rules that are developed by private enterprises and have no legal effect whatsoever?" Indeed, standards are voluntary: their non-binding nature is the source of their legitimacy and one of the reasons behind their wide implementation (Brunsson & Jacobsson, 2000). No national or international legal framework will oblige a manufacturer of electronic devices to use Bluetooth specifications in its products; but the markets will.

In a number of sectors, voluntary standards have become de facto rules for the industry (i.e. specifications for the Internet and Web), or metamorphosed into mandatory requirements, like it was the case for International Financial Reporting Standards in the EU (Chiapello & Medjad, 2009). In other sectors, standards guide stakeholders' behavior by enabling market access (Thorstensen, Weissinger, & Sun, 2015) or creating normative expectations (Werle, 2001a, b). It is the normative function of standards that provides a basis for compliance and incentivizes cooperation, which in turn increases network externalities (Delimatsis, 2018). Accountability of such private schemes is not achieved via formal legal proceedings, but by peer pressure and actors' reputation (Grant & Keohane, 2005). However, industry requirements may be "hardened" by explicit transportation to the national law, or by the means of a regulatory reference: the US Food and Drug Administration uses standards for medical device-communication development by IEEE as a guidance for device manufacturers (Schneiderman, 2015); building codes and procurement specifications often incorporate private standards (Abbott & Snidal, 2001); and compliance with harmonized European standards developed by private voluntary organizations presumes conformity with essential legal requirements formulated by the European Commission (EU Council Resolution, 1985). Following the series of European Court of Justice (CJEU) jurisprudence, such type of standards is also submitted to the constitutional principles of the European Union (EU), and their validity can be challenged in Courts by third parties (CJEU, 2012); (CJEU, 2016).

Despite standardization's strong resemblance to law-making, standards should not be regarded equivalent to law (Schepel, 2005). This does not mean, however, that standards do not have a legal value, which sometimes is even more far-reaching than the one of traditional regulation (Peters et al., 2009). Legal concerns associated with standardization activities go far beyond the evident antitrust and patent law violations and cover such issues as liability and basic human rights. For this reason, it is important that technical experts realize the extent to which their decisions may affect a broad range of stakeholders, and which consequences they may have for civil society.

3.3 Consensus and Openness Always Means a Race to the Bottom

"Why would I cooperate with companies from developing countries? Their safety requirements are so much lower than what we propose!" – an expert from downstream oil and gas industry once confessed to me his thoughts on multilateral cooperation in standards development. His preference for standardization platforms that are only comprised of Western stakeholders over their global counterparts was well substantiated. Especially when it comes to health and safety requirements, there is a limit to how much a party can compromise: the discussions on hormones in meat products that took place during the negations of Transatlantic Trade and Investment Partnership (TTIP) were a good example of this. I, however, find troublesome his rejection (and that of a number of experts from other industries) of consensus and openness as the main principles of standards development.

In technical standardization, consensus is often accompanied by openness; openness is preserved at the expense of time; and time is a crucial factor for R&D—intensive industries. The development of Bluetooth specifications slowed down once new players joined the Bluetooth SIG (Keil, 2002). The decision of a group of browser developers to move the formulation of Hypertext Markup Language (HTML) specifications from World Wide Web Consortium (W3C) to the Web Hypertext Application

Technology Working Group (WHATWG) was based on the formers' shift to Extensible Markup Language (XML) specifications, but also on the slow progress and lack of concrete results in W3C consensus process. As stated by one of the WHATWG leaders, "The Web is, and should be, driven by technical merit, not consensus" (Way, 2011).

Openness and consensus are well-established principles in the WTO framework covering standardization activities (WTO, 2000). Openness may refer both to the process and outcomes of standards development, and is strongly related to such elements as fairness, transparency and due process (Krechmer, 2006). Standards created in an open and transparent procedure are believed to have a wider scope of adoption but may also risk diminished returns on R&D investments (West, 2003). Although openness has multiple dimensions (West, 2007), it should not be confused with Open Source software, where the source code is made available free of charge to all users and implementers.

In turn, consensus is achieved when there is an "absence of sustained opposition to substantial issues," given that the views of all parties concerned are taken into account (ISO/IEC, 2004). The impact which consensus-building has on standards development processes would largely depend on the industry (Thorstensen et al., 2015). While reaching consensus proves challenging for stakeholders with conflicting interests (De Vries, Winter, & Willemse, 2017), studies demonstrated that in the longer run, consensus proves beneficial for standardization stakeholders (Simcoe 2012), "unlocks" new markets (Austin & Milner 2001); facilitates the diffusion of a standard (Werle, 2001a, b); and contributes to its legitimacy (Brunsson & Jacobsson, 2000). At the same time, consensus may not only lead to "race to the bottom" feared by Western companies, but undermine a standard's relevance to the market, or create a situation where weaker countries become bound by the decisions of developed States.

It remains a matter of standards bodies and groups to establish the processes and principles according to which their members cooperate. However, especially with the view of the potential regulatory impact of voluntary standards, it is reasonable to suggest that stakeholders affected by industry standards should be provided with opportunities to participate or at a minimum, to evaluate their establishment. Consensus and openness may not always be the most practical solutions, but they certainly contribute to preserving standardization global and inclusive.

3.4 The Need to Raise Awareness of Cross-sectoral Differences in Standardization

Building on actual conversations, the three examples illustrate the discrepancy of approaches to standardization activities across a number of sectors. Given that the increased interoperability and hyper-connectivity blur the distinction between standardization segments, misapprehensions amongst disciplines undermine cooperation and may cripple technical progress. When an industry owes its survival to the existence of other businesses, understanding of cross-sectoral concepts is crucial, since chances that industries will "borrow" each other's governance models are fairly strong. For instance, European telecommunications industry had to revisit its approach to patent licensing once the US-based Motorola has joined standardization of GSM technologies (Pelkmans, 2001).

Standard-setting integrates elements of information management, business science, political economy and law (De Vries, 2015). Largely driven by technical decision-making, standardization is not immune to political considerations (Spruyt, 2001). The rapid pace of globalization, together with the growing number of electronic devices and diffusion of Internet, does not only raise technical challenges of device interoperability and connectivity speed, but also concerns the fundamental issues of privacy and security: hence, standards have a profound impact on human life (see Cath & Floridi, 2016). For this reason, technical and technological standard-setting should take into account societal and legal aspects of private voluntary standards. Understanding of the role and value of such standards in global regulatory regimes is important for the consecutive questions of applicability of antitrust provision to standardization activities of standards organizations (CJEU 2010), liability of certification bodies to third parties (CJEU, 2017) and copyright of standards developed by (semi-) private institutions (CJEU, 2016). At the same time, it should be acknowledged that the role of law in private standardization activities is rather limited and may not overshadow voluntary nature of technical standards: the opposite may not only diminish effectiveness of industry-driven specifications, but open avenues for regulatory capture (Spruyt, 2001).

While the need to raise awareness among the variety of scientific approaches is widely recognized in standardization research community, more can and should be done to introduce multi-disciplinary methods in standardization education. Since the lack of awareness of cross-sectoral challenges constitutes one of the main obstacles for constructive dialogue amongst stakeholders, educating the current and future standardization community of professionals is of critical importance. In light of the conversations and interviews with experts, it is my strong belief that by familiarizing students and professionals with different aspects of standardization, we will prepare (future) experts for inevitable cross-sectoral cooperation and ensure increased quality and applicability of standards. The next section proposes a number of initiatives for current educational programs.

4 Education on Standardization: Ideas, Suggestions and Caveats

4.1 Limits of an Inter-disciplinary Approach

It goes without saying that, to fully grasp the concepts on which a sector is built, additional training, either scientific or technical, is required. Yet, combining a number of scientific disciplines without understanding their insights may be impractical, provide superficial results or even bear little relevance to the "real-world" situations (by analogy, Siems, 2009; Larouche, 2012). Educating legal and human scientists as engineers is often an unbearable task; likewise, engineers and computer scientists are not easily requalified to become lawyers or sociologists. And while some disciplines, such as law and economics, are commonly blended for education and research purposes, other interdisciplinary combinations are more challenging and would perhaps require years of additional training.

This contribution suggests, however, that such degree of inter- disciplinarity is not necessary in standardization education, unless reasonably required in professional setting. Similar to interdisciplinary research, interdisciplinary *education* can vary in levels of synthesis. In this regard, inter-disciplinarity may be perceived as a progression along different stages, where initial phases represent the acknowledgement and minimum awareness of other discipline(s), which later enables students to draw comparison between different scientific areas and to engage in the meaningful interdisciplinary dialogue and, eventually, to integrate the disciplines into a large whole (Larouche, 2012).

Given that the objective of introducing of multi- and inter-disciplinary approaches in standardization is to explain the way in which standards works and enhance the understanding of their effects on global society, how should the system on standardization education then be designed?

While learning is certainly a lifelong process, education can generally be divided in two phases: pre-qualification, or higher education, when students obtain the knowledge and information necessary to enter a particular profession, and post-qualification, when specialists follow additional training to strengthen their skills, acquire specific expertise or deepen their knowledge. During the first phase, (traditional) education is usually offered by universities and colleges, often on a full-time basis, while post-qualification courses are mostly short-term and highly-intensive programs taken by experts in parallel to their employment. While, following authors' experience, this submission mainly focuses on higher academic education, it also suggests how inter- and multi-disciplinary can be introduced in both phases of education on standardization, including non-traditional learning methods, such as online courses.

4.2 Pre-qualification

In recent years, college and university education was marked by gradual introduction of standardization-related topics into undergraduate and graduate curricula. The interplay between standardization and IP is discussed during management and digitalization classes; guidelines and requirements for food safety set by Codex Alimentarius is quite familiar amongst student of food management and trade economics; jurisprudence on ICT standardization represents the essence of antitrust and IP law courses. Yet, present curricula of most educational institutions do not adequately cover multi-disciplinary dimension of standardization and typically limit study material to the field in which students would obtain their (under) graduate degree. As a result, future engineers have little knowledge on societal aspects of standards, and future lawyers often fail to understand the technical rationale behind industry specifications, even when working in the ambit of economic law or law and technology.

Past experiences with (under) graduate educational programs suggest that students are motivated to follow standardization courses when they understand the pragmatic reasons behind standardization (Purcell & Kelly, 2003). Hence, students with no practical experience—or with a very limited academic background—may not be expected, in all reasonableness, to grasp the multifaceted aspects of standardization. For this reason, introduction of standardization courses for undergraduates may not be optimal: rather, students may benefit from one or two lectures within a course, which would enable them to understand the basics of standardization from the perspective of the main discipline of their study.

Take, for instance, bachelor courses in law and legal studies. When teaching the basics of European or national administrative law, differences between standards and legal documents, or governmental and private standardization, may be explained based on practical examples from students' daily lives, such as university guidelines for passing exams or national food safety requirements complied with by most cafeterias and bars. Likewise, during introductory courses in European competition law, standardization agreements may be used as an example of potentially collusive behavior under Article 101 of the Treaty of Functioning of the European Union (TFEU), and excessive royalty-rates for SEPs may as well illustrate abuse of dominance under 102 TFEU. This educational stage should not necessarily aim to provide a thorough analysis whether standards-related practices breach the relevant provisions of the EU or national law: rather, students should be able to understand the benefits and drawbacks of standardization, and the interplay between standards and regulation.

Inter- and multi-disciplinary should then be introduced in later education stages when students have already gained sufficient knowledge and understanding of a scientific field and are able to engage in a dialogue with students and experts from other areas. For specialized master programs, such as Law & Technology, Industrial Design or Technology Management, standardization can be introduced as an elective course, which would cover such topics as strategic standardization, antitrust risks, standardization governance, reduction of trade barriers, and business application of industry standards—based on students' needs and the envisaged purpose of the educational program. For less specialized degrees, standardization can be introduced within a context of a single course by delivering (a series of) guest lectures, where invited practitioners and experts from different disciplines would share their knowledge and experience in standards development and implementation, closing the lecture with a general discussion on standardization matters. Most importantly, key concepts of the disciplines should be explained in a professional, yet understandable language, and technical jargon should be either clarified or avoided. This would broaden students' horizon, but also allow them to view similar challenges from different perspectives and interact with experts from the field other than theirs.

Education about standardization may also be integrated in simulation activities where students are expected to demonstrate a high degree of independence in resolving semi-professional challenges, as well as ability to work in a team. For law students, these qualities are typically tested in moot courts or boardroom simulations: while dealing with a legal question arising from standardization practices (i.e. whether a private standard violates various provisions of WTO, antitrust or national IP law), students also engage in a wider discussion on standards that go beyond a single discipline. This approach proved to work in moot court and boardroom simulation exercises for bachelor students of Global Law program at Tilburg Law School, where by students were dealing with complex legal problems involving food-packaging standards, licensing policies of SEPs and standards' role in national legal systems.

4.3 Post-qualification

To target students in the final stage of their degree, or short after their graduation, summer schools or tailored standardization courses with invited experts would be a suitable option: these formats allow a specialized approach while also providing more room for multi-disciplinary discussions, moreover allowing students to expand their professional network. In this regard, courses covering standards-related topics may as well be a part of a summer schools' curriculum in specific fields, such as competition economics or IP law, but a separate short academic program on standardization is also much welcomed.

Specialized courses or training would be a preferred education method for professionals wishing to expand their insights on standardization to different domains. Excellent examples are: World Standards Cooperation (WSC) Academic Days, organized by ISO, IEC and ITU to promote dialogue and foster cooperation between standardization community and academics; StandarDays, a two-day session offered once a year by CEN and CENELEC; and ETSI Seminar held twice a year in ETSI Headquarters. Similar programs can be implemented by international and national standards organizations, each of them focusing on matters relevant for specific industry, country or region. Depending on their business needs, companies may also implement standardization trainings for employees: attending similar trainings on i.e. bribery and export controls constitute an obligatory part of newcomers' training in most of globally operating enterprises. Ultimately, law firms can offer tailored sessions covering legal aspects of technical standardization activities.

4.4 Online Learning Platforms

Similarly to many academic programs, multi-disciplinary education on standardization should not be limited to traditional education systems. The option to deliver standards education online or as web-based courses has been explored before but mainly in relation to materials offered on the websites of international standards bodies (i.e. Spivak & Kelly, 2003). Yet, more can be done to integrate standards and standardization in general courses on law, economics engineering or ICT offered by prominent universities by the means of internet learning platforms, such as Coursera, Udemy and Lynda.Com. A successful example from the past are the web-based lecture series "Standardization in companies and markets," a joint creation by the leading standardization scholars from Asian and European universities (Hesser & Siedersleben, 2007). Distance learning, although it often cannot replace traditional education methods, is an excellent tool for both students and professionals, and should be promoted by educational institutions.

5 Conclusion

Standards are on the verge of globalization and technological convergence: their importance for multilateral trade, security, quality and interconnectivity is well documented in media reports, government documents and academic literature. In the absence of its own scientific designation, standardization represents a patchwork of disciplines and is treated differently depending on the sector, professional setting and national legislation. While this contribution looks through the prism of law, it recognizes that similar misunderstandings may arise regarding other scientific fields but emphasizes that these misunderstandings would stem from lack of consideration for and knowledge on different standardization concepts. In the light of society's increased reliance on individual company-schemes or collective international schemes and the synthesis between different scientific and regulatory domains, traditional approaches to standardization became too narrow and may even lead to uninformed and imbalanced decisions. In this regard, call for action to bring awareness of cross-sectoral standardization is growing.

This contribution submits that such awareness can be achieved by introducing multi-disciplinary elements in education about standardization and by this means, prepare standards professionals who possess a myriad of skills relevant for handling standards-related activities. Multi-disciplinary research on standardization is on the rise, which also boosts hope for strengthening multi-disciplinary education about standardization. More specifically, this contribution welcomes introduction of standards-related topic in academic and professional law programs, as well as discussing legal elements of standardization in pre- and post-qualification programs in management, economics, engineering and other disciplines. Yet, it also acknowledges that standardization is a practical topic to teach and cautions that "over-ambition" in inter-and multi-disciplinary education may be detrimental to the overall purpose of educational programs. Nevertheless, the author sincerely hopes that shedding some light onto different aspects of standardization will benefit a broad community of scholars and practitioners, irrespective of their professional background.

In the story of the Tower of Babel, people accepted their punishment and abandoned the construction. Who knows, whether the Tower may still have been built had they learned the basic concepts of each other's language to the extent that they can collaborate and realize their ambitious dreams.

References

Abbott, K. W. (1989). Modern international relations theory: A prospectus for international lawyers. *Yale Journal of International Law, 14*(2), 335–411.

Abbott, K. W., & Snidal, D. (2001). International 'standards' and international governance. *Journal of European Public Policy, 8*(3), 345–370.

Austin, M. T., & Milner, H. V. (2001). Strategies of European standardization. *Journal of European Public Policy, 8*(3), 411–431.

Baron, J., & Pohlmann, T. (2015). *Mapping Standards to Patents Using Databases of Declared Standard-Essential Patents and Systems of Technological Classification*. Retrieved from http://www.law.northwestern.edu/research-faculty/searlecenter/innovationeconomics/documents/Baron_Pohlmann_Mapping_Standards.pdf.

Berman, P. S. (2005). From international law to law and globalization. *Columbia Journal of Transnational Law, 43*, 485–557.

Berman, P. S. (2012). *Global legal pluralism: A jurisprudence of law beyond borders*. New York: Cambridge University Press.

Biddle, B., White, A., & Woods, S. (2010). *How Many Standards in a Laptop? (And Other Empirical Questions)*. Retrieved from http://dx.doi.org/10.2139/ssrn.1619440.

Brunsson, N., Jacobsson, B., et al. (2000). *A world of standards*. New York: Oxford University Press.

Büthe, T. (2010). Engineering uncontestedness? The origins and institutional development of the International Electrotechnical Commission (IEC). *Business and Politics, 12*(3), 1–62.

Cath, C., & Floridi, L. (2016). The design of the internet's architecture by the Internet Engineering Task Force (IETF) and human rights. In *Science and Engineering Ethics*. https://doi.org/10.1007/s11948-016-9793-y.

Cassese, S. (2005). Administrative law without the state? The challenge of global regulation. *N.Y.U International Law and Politics, 37*, 663–694.

Cervigni, G., & Larouche, P. (2014). *Regulating Smart Metering in Europe: Technological, Economic and Legal Challenges*. Retrieved from http://www.cerre.eu/sites/cerre/files/140331_CERRE_SmartMetering_Final.pdf.

Chiapello, E., & Medjad, K. (2009). An unprecedented privatization of mandatory standard-setting: The case of European accounting policy. *Critical Perspectives on Accounting, 20*(4), 448–468.

Contreras, J. L. (2011). *An Empirical Study of the Effects of Ex Ante Licensing Disclosure Policies on the Development of Voluntary Technical Standards, conducted for the National Institute for*

Standards and Technology (NIST), US Department of Commerce. Retrieved from https://www. nist.gov/sites/default/files/nistgcr_11_934_empircalstudyofeffectsexantelicensing2011_0.pdf.

Contreras, J. L. (Ed.). (2018). *The Cambridge handbook of technical standardization law. Competition, antitrust and patents.* New York: Cambridge University Press.

Curran, P. D. (2003). Standard-setting organizations: Patents, price fixing, and per se legality. *The University of Chicago Law Review, 70,* 983–1009.

De Vries, H. J. (1997). Standardization—What's in a name? *Terminology—International Journal of Theoretical and Applied Issues in Specialized Communication, 4*(1), 55–83.

De Vries, H. J. (2002). Standardization education. ERS-2002-82-ORG. *ERIM Report Series Research in Management.* Retrieved from https://www.iso.org/sites/materials/initiatives-in-education/education_initiatives-higher-edu/educational_materials-detail-initiativese245.html? emid=21.

De Vries, H. J. (2003). Learning by example—A possible curriculum model for standardization education. *ISO Bulletin July 2003,* pp. 25–29.

De Vries, H. J., & Egyedi, T. M. (2007). Education about standardization—Recent findings. *International Journal for IT Standards & Standardization Research, 5*(2), 1–26.

De Vries, H. J. (2015). Standardization—A developing field of research. In P. Delimatsis (Ed.), *The law, economics and politics of international standardization* (pp. 19–41). New York: Cambridge University Press.

De Vries, H., Winter, B., & Willemse, H. (2017). Achieving consensus despite apposing stakes: A case of national input for an ISO standard on sustainable wood. *International Journal of Standardization Research, 15*(1), 29–47.

Delcamp, H., & Leiponen, A. (2014). Innovating standards through informal consortia: The case of wireless telecommunications. *International Journal of Industrial Organization Elsevier, 36C,* 36–47.

Delimatsis, P. (Ed.). (2015). *The law, economics and politics of international standardization.* New York: Cambridge University Press.

Delimatsis, P. (2018). Global Standard-Setting 2.0: How the WTO spotlights ISO and impacts the transnational standard-setting process. *Duke Journal of Comparative and International Law, 28,* 273–326.

European Commission. (2017). *Communication from the Commission to the European Parliament, the Council and the European Economic and Social Committee Setting Out the EU Approach to Standard Essential Patents,* COM, 712 final.

European Council. (1985). *Resolution on a New Approach to Technical Harmonization and Standards* (OJ C 136/1).

European Court of Justice. (2010). *Case T-432/05 EMC Development AB v European Commission.*

European Court of Justice. (2012). *Case C-171/11 Fra.bo SpA v Deutsche Vereinigung des Gas-und Wasserfaches.*

European Court of Justice. (2016). *Case C-613/14 James Elliott Construction Ltd v Irish Asphalt Ltd.*

European Court of Justice. (2017). *Case C-219/15, Elisabeth Schmitt v TÜV Rheinland LGA Products GmbH.*

Grant, R. W., & Keohane, R. O. (2005). Accountability and abuses of power in world politics. *American Political Science Review, 99*(1), 29–43.

Hesser, W. (2014). *Memorandum on Standardization in Higher Education in Europe.* Retrieved from https://www.iso.org/sites/edumaterials/hesser-memorandum.pdf.

Hesser, W., & Czaya, A. (1999). Standardization as a subject of study in higher education. A vision. *ISO Bulletin June 1999,* pp. 6–12.

Hesser, W., & Siedersleben, W. (2007). Standardization goes East. The European—Asian academic network. International and multi-media based. *ISO Focus November 2007,* pp. 21–24.

International Organization for Standardization and International Electrotechnical Commission. (2004). *Guide 2: Standardization and Related Activities—General Vocabulary.*

The Need for Multi-disciplinary Education About Standardization

Keil, T. (2002). De-facto standardization through alliances—lessons from Bluetooth. *Telecommunications Policy, 26*(3), 205–2013.

Krechmer. K. (2006). Open Standards Requirements. *International Journal of IT Standards and Standardization Research, vol. 4 (1).*

Krechmer, K. (2007). Teaching standards to Engineers. *International Journal of IT Standards and Standardization Research, Vol. 5*(2).

Lane, C. (1997). The social regulation of inter-firm relations in Britain and Germany: Market rules, legal norms and technical standards. *Cambridge Journal of Economics, 21*(2), 197–215.

Larouche, P. (2012). A vision of global legal scholarship. *TILEC Discussion Paper* (No. 2012-034). Retrieved from https://pure.uvt.nl/ws/files/1461065/2012_034_1_.pdf.

Lundell, B., Gamalielsson, J., & Katz, A. (2015). On implementation of open standards in software: To what extent can ISO standards be implemented in open source software? *International Journal of Standardization Research, Vol. 13*(1).

O'Sullivan, E., & Brévignon-Dodin, L. (2012). *Role of standardization in supporting emerging technologies. A Study for the Department of Business, Innovation & Skills (BIS) and the British Standards Institution (BSI).* Retrieved from https://www.ifm.eng.cam.ac.uk/uploads/Resources/Reports/OSullivan_Dodin_Role_of_Standardisation_June_2012__2_.pdf.

Pelkmans, J. (2001). The GSM standard: Explaining a success story. *Journal of European Public Policy, 8*(3), 432–453.

Peters, A., Koechlin, L., Förster, T., & Fenner Zinkernagel, G. (Eds.). (2009). *Non-state actors as standard setters.* New York: Cambridge University Press.

Posner, R. A. (1987). The decline of law as an autonomous discipline. *Harvard Law Review, 100,* 1962–1987.

Purcell, D. E., & Kelly, W. E. (2003). Adding value to a standards education: Lessons learned from a strategic standardization course. *ISO Bulletin July 2003*, pp. 33–34.

Russel, A. L. (2006). Rough consensus and running code' and the internet-OSI standards war. *IEEE Annals of the History of Computing*, pp. 48–61. Retrieved from https://pdfs.semanticscholar.org/9ffa/d637b841df9e1904aea2265d0a88fd855d58.pdf.

Schepel, H. (2005). *The constitution of private governance: Product standards in the regulation of integrating markets.* Oxford: Hart Publishing.

Schneiderman, R. (2015). *Modern Standardization: Case Studies at the Crossroads of Technology, Economics and Politics.* Standards Information Network.

Siems, M. M. (2009). The taxonomy of interdisciplinary legal research: Finding the way out of the desert. *Journal of Commonwealth Law and Legal Education, 7,* 5–17.

Simcoe, T. (2012). Standard setting committees: Consensus governance for shared technology platforms. *American Economic Review, 102*(1), 305–336.

Spivak, S. M., & Kelly, W. E. (2003). Introduce strategic standardization concepts during higher educational studies…and reap the benefits! *ISO Bulleting, 2003,* 22–24.

Spruyt, H. (2001). The supply and demand of governance in standard-setting: Insights from the past. *Journal of European Public Policy, 8*(3), 371–391.

Thorstensen, V., Weissinger, R., & Sun X. (2015). *Private Standards—Implications for Trade, Development, and Governance.* Retrieved from http://e15initiative.org/wp-content/uploads/2015/07/E15-Regulatory-Thorstensen-et-al.-final.pdf.

Way, J. (2011). *A Brief History of HTML5.* Retrieved from https://code.tutsplus.com/articles/a-brief-history-of-html5-net-23064.

Werle, R. (2001a). Institutional aspects of standardization—Jurisdictional conflicts and the choice of standardization organizations. *Journal of European Public Policy, 8*(3), 392–410.

Werle, R. (2001*). Standards in the international telecommunications regime.* HWWA Discussion Paper 157, SSN 1616-4814. Retrieved from https://www.econstor.eu/bitstream/10419/19394/1/157.pdf.

West, J. (2003). How open is open enough? Melding proprietary and open source platform strategies. *Research Policy, 32*(7), 1259–1285.

West, J. (2007). The economic realities of open standards: Black, white and many shades of gray. In S. Grrenstein & V. Stango (Eds.), *Standards and public policy* (pp. 87–122). Cambridge: Cambridge University Press.

World Trade Organization. (2017). *Russian Federation—Measures on the Importation of Live Pigs, Pork and Other Pig Products From the European Union*. Report of the Appellate Body, adopted on February 23, 2017. WT/DS475/AB/R.

World Trade Organization. (2000). *Annex 4: Decision on Principles for the Development of International Standards, Guides and Recommendations with Relation to Articles 2, 5 and Annex 3 of the TBT Agreement* (G/TBT/9.). Issued on November 20, 2000.

Wiegmann, P. M., De Vries, H. J., & Blind, K. (2017). Multi-mode standardisation: A critical review and a research agenda. *Research Policy, 46*(9), 1370–1386.

Wu, X. (2017). *Interplay between Patents and Standards in the Information and Communication Technology (ICT) Sector and its Relevance to the Implementation of the WTO Agreements*. WTO Working Paper ERSD – 2017-08 (2017). Retrieved from https://www.wto.org/english/res_e/reser_e/ersd201708_e.pdf.

Olia Kanevskaia is a Ph.D. Candidate at Tilburg Law and Economics Center (TILEC) and Tilburg Law School, the Netherlands. Her doctoral research focuses on the institutional aspects of technological standardization and the governance of industry-driven standards development bodies. She holds a master's degree in International and European Public Law from Tilburg University (2014). During her doctoral study, Olia was a graduate intern at the Trade and Environment Division of the WTO and a Visiting Fellow at the Center on Law, Business and Economics of Northwestern University. She previously worked for Eurojust, Europol and TNT Legal Affairs/Sanctions Team.

Learning in Communities of Standardisation Professionals

Henk J. de Vries, Jeroen Trietsch and Paul M. Wiegmann

1 Standardisation—A Discipline?

People working in a certain area get education and in many cases are members of a professional society. Typically, such a society enables its members to share knowledge and experience, and it contributes to further development of the field. One may wonder to which extent such professional societies do or should function for the field of standardisation. This paper aims to answer this question.

Verman (1973), the former director of the Indian Standard Institution, described standardisation as a discipline for which specific knowledge is relevant and concludes (p. 441):

> It is a prerequisite that a full-fledged discipline of standardization be created, that standardization ceases to be an activity ancillary to engineering, industrial engineering, industrial management, economic planning or whatever else. This would need intensive thinking and

---

Note: An earlier version of this chapter was presented at the 11th International Conference "Standardization, Protypes and Quality: A Means of Balkan Countries' Collaboration" in Belgrade, 2014, and published in its proceedings (de Vries, Trietsch, & Wiegmann, 2014). A shortened version was published in the journal of American and Canadian standards professionals *Standards Engineering* (de Vries, 2015).

H. J. de Vries (✉)
Rotterdam School of Management, Erasmus University, Rotterdam, The Netherlands
e-mail: hvries@rsm.nl

Faculty of Technology, Policy and Management, Delft University of Technology, Delft, The Netherlands

J. Trietsch
Knowledge Network for Continuous Improvement, 's-Hertogenbosch, The Netherlands
e-mail: jeroen.trietsch@gmail.com

P. M. Wiegmann
Technical University of Eindhoven, Eindhoven, The Netherlands
e-mail: p.m.wiegmann@tue.nl

© Springer Nature Switzerland AG 2020
S. O. Idowu et al. (eds.), *Sustainable Development*, CSR, Sustainability, Ethics & Governance, https://doi.org/10.1007/978-3-030-28715-3_11

extensive organizational work. Many more men of high calibre would need to be attracted to adopt standardization as a profession. Those within the profession will need to reorient their thinking. Fortunately, it is a basic principle of standardization to work and think collectively and cooperatively. This feature would have to be exploited to the fullest possible extent by pooling all the available resources – material as well as mental. At the same time, individual thinking and original contributions to the advancement of the discipline should be encouraged to the fullest. Both individual and cooperative efforts are essential. When the discipline emerges from its present status and wins for itself all-round recognition, the time will be ripe to explore and promote the numerous possibilities of the future in the context of the world that is still to come.

Kuhn (1970) describes the development of scientific disciplines and relates this to the creation of communities of practitioners. de Vries (1999, Sect. 1.1.3) applies this to standardisation. Combining these studies we can conclude that standardisation needs such communities of practitioners, both at the national and at the international level, and that these need to be connected to the domains of scientific research and education.

Such communities have indeed been formed. Several countries established national standards users organisations such as the Standards Engineering Society in the United States and Canada. Europe has a scientific community: the European Academy for Standardisation EURAS, which is also open for professionals from industry, standards bodies, and consultancy firms. The Balkan Committee for Standardisation, Protypes and Quality unites scientific people and people from organisations in the field of standardisation and conformity assessment. Bulgaria, Greece and South Korea show examples of national organisations in which both scientists and standardisation practitioners participate. Recently the China Jiliang University established the Belt and Road University Alliance of Standardization (B&RUAS), aiming at education and research in the field of standarisation. This initiative unites 106 universities, many from China but also from other countries all over the world. The International Cooperation on Education about Standardization (ICES) is an informal international network for sharing knowledge and experiences related to education about standardisation.

To conclude, some initiatives for sharing standardisation knowledge are in place though this seems far from 'mature'. This chapter therefore first highlights some academic findings on structures for sharing knowledge, then describes some of these in the field of standardisation, and finally discusses a possible way to proceed.

2 Structures for Sharing Knowledge and Networking

Academic literature distinguishes between two settings in which people can network and exchange knowledge: Communities of Practice (CoPs) and Networks of Practice (NoPs).

CoPs are networks of people or organisational units that share common interests and communicate about these interests in some form to exchange experience about their topic of interest and generate new ideas (Wenger & Snyder, 2000). Very different types of networks are covered by the term CoP. Wenger, McDermott, and Snyder (2002) name seven factors that characterise different CoPs (their boundaries; size; lifespan; members' location; types of members; formal organisation of the CoP and the motivation to start the CoP). In the context of communities of standards professionals, some of these characteristics are especially interesting as they show the large variety of communities that could potentially exist to support professionals in the area. For example, the boundary of such a CoP can be limited to a specific part of a company, several business units or across different organisations (Allee, 2000; Wenger et al., 2002). Members in a CoP can be located closely together or geographically dispersed. A CoP can also be organised as a formal entity (such as a non-profit organisation) or an informal, not institutionalised gathering of like-minded people. When CoPs fail this can have several reasons (Probst & Borzillo, 2008; Roberts, 2006): There may be a lack of core members who hold the network together, a low level of interaction between members and a large share of members who do not learn from others' expertise or do not value the CoP highly in their daily work. In order to avoid a CoP's failure, Probst and Borzillo (2008) and Roberts (2006) list critical success factors (focus on a central topic that is valuable for all members; presence of a central, well-respected coordinator; sufficient amount and time invested by participants; building on members' values to support knowledge sharing; involvement of well-respected leaders; different activities to support contact between members; enthusiastic supporters of the CoP must attract new members for whom it should be easy to join; contributing must be easy; relevant issues should be discussed in the CoP). However, despite the existence of such success factors it is difficult to measure success of a CoP. Characteristics that may be associated with success in one CoP may be seen as failure in another, success seems to be mainly associated to how members perceive their CoP (Damianos & Holtzblatt, 2010).

NoPs are more formal networks that link people who may never have met but are working in the same field (Brown & Duguid, 2000). In these systems, members do not interact directly but information is transmitted via the indirect links that are created by the network which means that little new knowledge is created in them (ibid.). This also means that a network centre is needed which integrates knowledge from different parts of the network (Andriani, Atkinson, Bowden, & Hall, 2004). Andriani et al. (2004) find three ways in which NoPs differ from CoPs: Knowledge in the network is created by the individual members; NoPs are very modular, and Nonaka and Reinmoeller's (1998) SECI model does not explain knowledge creation in NoPs.

Organisations that allow professionals in standardisation to develop their professional skills may therefore resemble CoPs or NoPs, depending on their set-up. If they aim to provide a forum in which professionals can exchange experience and commonly develop solutions to new challenges, they are more likely to resemble CoPs, whereas organisations that aim to provide education and training may resemble NoPs more.

3 International Academic Society: EURAS

Apart from the Belt and Road University Alliance of Standardization (B&RUAS), that still needs to establish its modus of operations, and the Balkan Committee of Standardization, Protypes and Quality, the European Academy for Standardisation EURAS is the only international community of academic researchers in the field of standardisation. EURAS was founded in Hamburg, Germany, in 1993 by researchers from various academic fields. The foundation of EURAS was prompted by a common desire to promote and achieve progress in the academic treatment of standardization, involving the widest possible range of disciplines. EURAS' objective is to promote research, education and publications in the field of standardisation. The society's activities focus on promoting standardisation research, a critical evaluation thereof in the interest of scientific education, improving opportunities to publish research results, and supporting the development and professionalisation of education about standardisation.

EURAS does not address technical research related to the contents of specific standards. Such pre-normative research is definitely important but EURAS focuses on research about standardisation as a phenomenon, in order to better understand it. Such better understanding may form the basis of improvements of the system, and for education. A multitude of disciplines is needed for this understanding, such as economics, business, engineering, sociology, psychology, public policy, law, history and information sciences. EURAS wants to promote interdisciplinary research on standards of all kinds and in all fields. Its main activity is the yearly conference. The conference provides opportunities for knowledge exchange and feedback among peers, primarily from the academic world but also from industry and standards bodies. The informal contacts during the conference may lead to research cooperation in the future.

The second field of activity of EURAS is in stimulating education about standardisation. EURAS prepared a *White Paper on Education about Standardization in Europe* (Hesser & de Vries, 2011) that formed the basis for current activities as part of the *Joint Initiative on Standardisation* in which the European Commission and many other European and national stakeholders cooperate to strengthen standardisation in Europe.

4 National Professional Societies: Standards Users Organisations

The oldest organisations for generating and exchanging standardisation knowledge, sometimes more than 40 years old, are the national standards users organisations. Traditional membership of these organisations consists of standards officers in larger companies. However, their positions are in jeopardy or have disappeared due to several reasons such as the following:

Learning in Communities of Standardisation Professionals 183

- Their work is service work with limited possibilities for automation. Other functions within companies, in particular production, can profit much more from automation. This has made standardisation relatively more expensive.
- Their work is central administrative staff work. In recent years it is business fashion to de-staff administration and leave it to the line organisation to take up traditional administrative tasks. This applies to other former central administrative functions such as human resource management and quality management but also to the responsibility for standardisation. As a result, systematic attention for and knowledge of standardisation disappears but who cares?
- This 'who cares?' relates to the company's incentive structure. The Anglo-Saxon short-term shareholder thinking tends to ignore standardisation because standardisation requires by definition a medium- to longer-term investment which can be skipped in the short term.
- Company management and business consultants lack knowledge about the strategic importance of standardisation—out of sight, out of mind.
- Practitioners lack knowledge and skills to do a good job and as a result often do a poor job, related to technical documentation rather than company strategy.[1]
- As a result the tasks may be given to relatively older people—experienced and conscientious but maybe less equipped for more dynamic and prestigious responsibilities.
- Lack of recognised education with related academic and professional qualifications hinders the recognition of the professional as well as the profession.

The lack of education relates to the lack of manifest demand—the request for standardisation education should come from industry but they lack the knowledge and vision to ask for it—a typical chicken-and-egg situation.

5 International Professional Society: IFAN

The organisation of standards professionals at the international level is the International Federation of Standards Users IFAN. Its members include national standards users organisations and some companies. 'The International Federation of Standards Users (IFAN) was born in 1974 as an ambassador of the standards users' interests in the international arena, long before globalization was heard of, and at a time when competition was less fierce than today, and trade barriers and protectionism were still accepted practice in certain areas' (IFAN, 2014). 'The International Federation of Standards Users is an independent, non profit-making international association of

---

[1] For instance, in the case of the task of industry participants in international standardisation, research has revealed more than 100 factors that contribute to successful participation in international standardisation committees (Brons, 2007). Most participants in international standardisation are not aware of such factors. They spend several days or weeks a year in standardisation activities and have the impression that they do a good job but are not aware that their efforts could be much more effective.

national organizations for the application of standards, companies, professional and trade associations, and governmental agencies, concerned with the use of standards. (…) The objectives of IFAN, inter alia, include:

- To promote uniform implementation of standards without deviation and develop solutions to standards users' problems without itself drawing up standards;
- To consolidate standards users' interests and views on all aspects of standardization and conformity assessment, and to cooperate with international and regional standardizing bodies (e.g. ISO, IEC, ITU, CEN, CENELEC, COPANT, etc.) in order to communicate user views to these organizations;
- To promote networking in the field of international standardization and conformity assessment'(IFAN, 2018).

IFAN is the only general organisation for professional standards users at the international level. It obtained recognition by formal standardisation organisations at the global level (ISO and IEC) and at the regional (European) level (CEN and CENELEC) as an organisation prepared to express the views of standards users and several of its members have a similar recognition at the national level. Indeed, IFAN and its national members have formulated user needs and serve as a (user) focus group for standards bodies. This impacts standardisation practice, for the benefit of users. Additionally, the national members serve as a platform for exchange of experiences between standardisation professionals.

However, IFAN faces a decline in membership. The number of countries with a national standards users organisation has decreased and if there is one, the value of membership of IFAN is not self-evident. This decrease in the number of standards users organisations applies to Europe, meanwhile an increase can be observed in Asia (Indonesia, South Korea). In addition there has been a decline in IFAN's corporate membership.

Another issue is that the terms 'standard user' is confusing, most people on the globe are standards users without being aware of it. At the international level, Copolco (ISO's Committee on Consumer Policy) and Consumers International try to give them a voice. Seen in this way, IFAN and its national members unite standards professionals rather than standards users but the number of people involved is only a fraction of the entire number of standards professionals.

6 Standards Professionals

The number of participants in standardisation committees globally is several hundreds of thousands, the number of people involved in standards implementation and use is millions. So potentially there is an enormous 'professional community'. Moreover, this community is increasing further due to the increasing importance of standards. Apart from 'traditional' reasons for this increase such as globalisation, the extension of standardisation to service areas and the increasing importance of

Learning in Communities of Standardisation Professionals 185

conformity assessment and related regulatory and trade issues, there are some more recent reasons such as the following:

- The increased awareness among politicians and government officials at the national (e.g., China, Germany, Malaysia) and regional (APEC, EU) level that standards are needed for innovation.
- This innovation more and more concerns complex systems consisting of integrated products and services rather than single products. Then standards have to provide stable interfaces between elements of these systems.
- Increased attention for societal issues creates a need for standards, often related to conformity assessment, and brings standards on the agenda of other organisations than just companies, such as NGOs.

All these people who spend a substantial share of their professional time to standards or standardisation could profit from better knowledge about standards and standardisation. Indeed, all over the world but in particular in Asia we can observe initiatives to get more attention to standards and standardisation in formal education at different levels (reaching from primary schools to universities) and in post-formal education and training. Different stakeholders need well-educated standards professionals:

- Notwithstanding the decrease in number of standards officers in companies, their importance has not decreased. On the contrary, they are even more important now. However, then the tasks should be carried out in a fully professional way, including scientific underpinning, and be directly related to company policy.
- Trade associations, governments and NGOs need to be triggered by knowledgeable standardisation experts to enable them to better relate standardisation to strategic goals, for the benefit of business and society.
- Standards bodies need knowledgeable standardisation experts to further professionalise and improve their services.
- Because standardisation and conformity assessment are getting more important while at the same moment companies are reluctant to appoint standards officers, we may expect that the number of standards consultants and related professions will grow.

Such full-time standardisers would benefit from a professional community. These professionals still can be found in some big companies, but also in standards bodies, as well in conformity assessment organisations, some consultancy firms, governments, and academia. This diverse group of full-time standardisers could meet at the national level to share knowledge and experience and further develop the discipline, and the national organisations can be united in an international one, also to share knowledge and experience and further develop the discipline. Apparently, if IFAN and its national members were not there already, they should be established. But maybe they should modernise their functioning. The South-Korean Society for Standards and Standardization can serve as a benchmark for such a diverse professional organisation at the national level. Its membership includes university professors and standards experts from industry, government, standards bodies and organisations in the field of conformity assessment. This provides an excellent opportunity to bridge the gap between academic and non-academic people active in the standardisation field.

7 Communities per Standardisation Task?

The vast majority of standards users are 'part-time standardisers' and they perform one or a few standards-related tasks. They never would become member of a standards users organisation but might nevertheless benefit from support for their task and from sharing experiences with other people performing a similar task. Therefore, in order to serve the majority of standards users it would be better to organise communities per task. Modern media provide opportunities to do so but this will not be sufficient. Standardisation tasks include (Simons & de Vries, 2002, p. 185):

- Signalling developments and trends relevant for standards and standardisation;
- Strategic analysis;
- Priority setting;
- Evaluation;
- Operational signalling;
- Operational analysis;
- Developing standards;
- Implementing standards;
- Assortment reduction and management;
- Participation in external standardisation;
- Providing information about standards;
- Managing the collection of company standards;
- Auditing standards use.

Typically the support for such a task may include the following:

- Providing a (face to face and/or electronic) platform for sharing information and experiences;
- Guidance documents on how to perform the tasks (including videos, apps and serious games);
- Developing standards for the profession[2];
- Developing job profiles;
- Education;
- Training;
- Conformity assessment for employees (diplomas, certification);
- Conformity assessment of accomplishments (audits, tests);
- Conformity assessment of education/training (inspection, certification, accreditation);

---

[2]Here a Korean initiative can be mentioned to develop an ISO International Workshop Agreement for "Competence Requirements for Standards Professionals in Companies and Organizations".

Learning in Communities of Standardisation Professionals

- Research (on how the task is being performed, impacts, improvements, innovation);
- Influencing service suppliers on behalf of users;
- Promoting the profession.

In the case of widely used standards (e.g., ISO 9001), support for standardisation tasks may also be standard-specific. Indeed, there are some standard-specific on-line communities. Support per task may be organised both at the national and at the international level.

8 Conclusion and Discussion

To conclude, the field of standards and standardisation deserves professional communities, both at the national and at the international level. The core could be communities for full-timers and these could initiate related communities per standardisation task.

Notably IFAN and its national members are in the position to take initiatives in this direction. Or should it be ICES, maybe in co-operation with IFAN? Anyhow, no matter which party jumps at this opportunity, it will require resources. These probably can stem from suppliers of commercial support for the standardisation tasks rather than from users. Potential sponsors include standards bodies at the national, regional and global level, conformity assessment organisations, national governments, and APEC, the EU and similar regional organisations.

A dilemma is whether such a 'modern' professional society at the national or international level should also take up commercial activities like education, training or curriculum certification. A benchmark in this sense, in the related field of quality management, is the American Society for Quality ASQ. Such activities could generate income and would give the society a more stable and powerful position but may affect independence.

References

Allee, V. (2000). Knowledge networks and communities of practice. *OD Practitioner, 32*(4), 1–15.

Andriani, P., Atkinson, G., Bowden, A., & Hall, R. (2004). Developing knowledge. In *Developing knowledge in "Networks of Practice"*. Singapore: World Scientific.

Brons, T. F. (2007). *Effective participation in formal standardization: A multinational perspective.* Rotterdam, The Netherlands: Rotterdam School of Management.

Brown, J. S., & Duguid, P. (2000). The social life of information. Boston, MA: Harvard Business School Press.

Damianos, L. E., & Holtzblatt, L. J. (2010). *Measuring community success: One size does not fit all.* The MITRE Corporation.

de Vries, H. J. (1999). *Standardization—A business approach to the role of National Standardization Organizations.* Dordrecht: Kluwer Academic Publishers.

de Vries, H. J. (2015). Building a community of standards professionals. *Standards Engineering, 67*(1), 8–11.

de Vries, H. J., Trietsch, J., & Wiegmann, P. (2014). Towards a community of standardisation professionals. In I. Mijatović, J. Filipovic, & A. Horvat (Eds.), *Proceedings 11th International Conference "Standardization, Protypes and Quality: A Means of Balkan Countries' Collaboration"* (pp. 13–20). Belgrade: University of Belgrade, Faculty of Organizational Sciences.

Hesser, W., & de Vries, H. J. (2011). *White paper academic standardisation education in Europe.* Hamburg/Rotterdam: European Academy for Standardisation.

IFAN. (2014). *IFAN 40th anniversary.* Geneva: IFAN.

IFAN. (2018). *IFAN in brief.* Geneva: IFAN. Retrieved 2018-06-02 from http://www.ifan.org/About-ifan.html.

Kuhn, T. S. (1970). *The structure of scientific revolutions.* Chicago: Chicago University Press.

Nonaka, I., & Reinmoeller, P. (1998). The "ART" of knowledge: Systems to capitalize on market knowledge. *European Management Journal, 16,* 673–684.

Probst, G., & Borzillo, S. (2008). Why communities of practice succeed and why they fail. *European Management Journal, 26*(5), 335–347.

Roberts, J. (2006). Limits to communities of practice. *Journal of Management Studies, 43*(3), 623–639.

Simons, C. A. J., & de Vries, H. J. (2002). *Standaard of maatwerk - Bedrijfskeuzes tussen uniformiteit en verscheidenheid.* Schoonhoven, The Netherlands: Academic Service.

Verman, L. C. (1973). *Standardization—A new discipline.* Hamden, CT: The Shoe String Press/Archon Books.

Wenger, E. C., McDermott, R., & Snyder, W. M. (2002). *Cultivating communities of practice.* Boston, MA: Harvard Business School Publishing.

Wenger, E. C., & Snyder, W. M. (2000). Communities of practice: The organizational frontier. *Harvard Business Review, 78*(1), 139–145.

Other Sources Used

Biesheuvel, J., Verkuyl, J., & de Vries, H. J. (1993). *Normalisatie in de metalektro, van beroeps-naar taakgericht werken.* 's-Hertogenbosch, The Netherlands: Nehem.

de Vries, H. J., Blind, K., Mangelsdorf, A., Verheul, H., & van der Zwan, J. (2009). *SME access to European standardization—Enabling small and medium-sized enterprises to achieve greater benefit from standards and from involvement in standardization.* Brussels: CEN and CENELEC.

de Vries, H. J., & Egyedi, T. M. (2007). Education about standardization—Recent findings. *International Journal for IT Standards and Standardization Research, 5*(2), 1–16.

Kurokawa, T. (2005) Developing human resources for international standards. *Science & Technology Trends—Quarterly Review, 5*(17), 1–10.

Henk J. de Vries (1957) is Professor of Standardisation Management at the Rotterdam School of Management, Erasmus University in Rotterdam, The Netherlands, Department of Technology and Operations Management, Section Innovation Management, and Visiting Professor at Delft University of Technology, Faculty of Technology, Policy and Management, Department of Values, Technology and Innovation, Section Economics of Technology and Innovation. His education and research focus on standardisation from a business point of view. From 1994 until 2003, Henk worked with NEN, Netherlands Standardization Institute, in several jobs, being responsible for R&D during the last period. Since 1994, he has an appointment at the Erasmus University's School of Management and since 2004, he has been working full-time at this university. Henk is (co-)author of more than 380 publications on standardisation, including several books.

See www.rsm.nl/people/henk-de-vries. In 2009, the International Organization for Standardization ISO awarded his education about standardisation as best in the world. Henk is President of the European Academy for Standardisation EURAS.

Jeroen Trietsch is Senior Supply Chain & Operations Consultant at EY, Amsterdam. His work focuses on helping manufacturing and services organisations becoming more successful by delivering their product in a more sustainable, faster and cheaper way. Jeroen is the founder of a Community of Practice for Continuous Improvement professionals in The Netherlands.

Paul M. Wiegmann is Assistant Professor at the Technical University of Eindhoven. His research addresses the dynamics that occur at the intersection between standardisation and innovation. Paul has (co-)authored publications in *Research Policy* and the *Handbook of Innovation and Standards*, as well as the recent book *Managing Innovation and Standards: A Case in the European Heating Industry* which shows how companies can address standards affecting their innovations.

Teaching Standardization to Generation Z-Learning Outcomes Define Teaching Methods

Ivana Mijatovic

1 Introduction

Standardization is neither a science nor even a generally accepted academic discipline. Standardization still might be described as a "new, emerging and multipoint discipline in which applied science, technology, industry and economics play extremely important parts" (Verman, 1973). Proponents of education about standardization claim that standardization needs to find its place in formal higher education and lifelong learning globally in order to create future professionals who can use their knowledge and skills related to standards and standardization for technology transfer, economic growth, global sustainability or in the case of latecomer countries for strategies of "accept and learn" or of (technology) "catch up" (Choung, Ji, & Tahir, 2011). Unterhalter and Carpentier (2010) consider "global higher education as a space for understanding the promises and pressures associated with competing demands for global economic growth, equity, sustainability, and democracy".

Almost all standards development organizations (SDOs) and many other international organizations call for more content about standardization in higher education curricula and for more activities to support education about standardization. In 1970, the UNECE Government Officials Responsible for Standardization Policies (the predecessor of WP. 6) developed Recommendation I, which urged governments to include standardization in the curricula of educational institutions. Recently several events and initiatives important for education about standardization took place. Based on EU—Joint Initiative on Standardization—Action 3: Programmes for education in standardization, on May 16th 2017, European Commission invited university professors from 22 European Countries, Serbia and Turkey to one day discussion entitled: "European Commission meets with Academia: Standardization in the 21st

---

I. Mijatovic (✉)
Faculty of Organizational Sciences, University of Belgrade, Belgrade, Serbia
e-mail: ivana.mijatovic@fon.bg.ac.rs; ivanamt@fon.bg.ac.rs

© Springer Nature Switzerland AG 2020
S. O. Idowu et al. (eds.), *Sustainable Development*, CSR, Sustainability,
Ethics & Governance, https://doi.org/10.1007/978-3-030-28715-3_12

century: societal, economic and educational aspects". More than 80 professors were there.

The topic of the Annual meeting of International Cooperation for Education about Standardization (ICES) 2017, held at Northwestern University, Pritzger School of Law, specifically focused on skill requirements and pedagogy in education about standardization. On the May 20th, 2018, at China Jiliang University in Hangzhou China, The Belt and the Road University Alliance for Standardization Education and Academics (B&RUAS, UAS) was established by 106 universities from 30 countries. The World Standards Cooperation (WSC) academic day "Leveraging Internet-based technologies to teach standardization" held as a back to back event with the ICES 2018 on July 5th in Yogyakarta, Indonesia, was dedicated to reviewing the potential of Internet-based technologies in teaching, training and education about standardization. The experience of the IEEE in the development and usage of Internet-based technologies and MOOCs in teaching and education about standardization was presented. The second European Commission meeting with academia was held in Sophia Antipolis on October 4th and 5th 2018 and hosted by ETSI; there the first version of the teaching materials for education about ICT standardization was presented.

Many initiatives have been taken in order to raise awareness of education about standardization but it seems that they are isolated from one another. More clearly, many initiatives about education about standardization, are starting from scratch and do not take into account what has already been done in that context. It seems that still focus of many interested parties in education about standardization is only on what should be a content of standardization course. Many, especially international and European SDOs have installed repositories of teaching materials and textbooks. Those efforts are highly valuable, but up to now, it seems that those initiatives are not widely accepted by the academic community. One of the reasons might be that contents are one-sided, some are overly based on promotion and not suitable for academic use.

Regular education courses have been much more actively supported and coordinated in countries such as Korea and China and less in Europe (see more in Choi & de Vries, 2013), however transferring good practices among higher education institutions (HEIs) are specific. Generally, educational systems are different around the globe; universities are of different quality and their functioning might be based on many different ways. In those contexts, varieties can be substantial even in one country and region or among the schools of the same university. "There is no universally valid formula for how a university must be, and it is the universal truth that universities must be different" (Greisler, 2013, p. 17). There is no single excellence model: "Europe needs a wide diversity of higher education institutions, and each must pursue excellence in line with its mission and strategic priorities" (European Commission, 2011, p. 2). Interest for education about standardization is growing; however, mutual action of several stakeholders is needed to overcome barriers (see more in de Vries, 2015).

Based on the framework for standardization education, elaborated in a study of Choi and de Vries (2011), which "presents an applicable combination of target groups (who), appropriate learning objectives (why), probable program operators (where), prospective contents modules (what), and preferred teaching methods (how)", this paper aims to provide more insight in specific aspects of teaching about standardization. The main research question in this paper is how to better facilitate a new generation of students—generation Z students—to learn about standardization.

It seems that success of many initiatives and action related to education about standardization actions is limited due insufficient understanding of higher education context, in the next paragraph three missions of universities will be explained. Higher education is changing and many higher education institutions (HEIs) are on their crossroads due to the increased need for internationalization and facts that the labor market is moving faster than higher education can respond. On the other side, universities are facing Generation Z with their unique values, preferences and learning motivation. In the next part of the chapter aspects of internationalization, roles of nontraditional providers of higher education and Generation Z will be elaborated. In the third part of the paper were elaborated what kind of teaching methods are more appropriate for generation Z and how teaching methods can be connected in learning outcomes related to standardization. After, this part conclusion remarks and reference were given.

2 Higher Education Is Changing

2.1 Three Missions of Universities

Universities have three complementary missions: teaching, research, and knowledge transfer to society (Montesinos, Carot, Martinez, & Mora, 2008). Teaching, research, and knowledge transfer to society sometimes do not have equal focus in all schools—some faculties and schools are more focused on teaching, some on research and some are dominantly oriented on providing practical knowledge (Light, Cox, & Calkins, 2009, pp. 38–43). Majority of the best universities at the global level are research universities and teaching and research are interconnected.

Even though, Von Humboldt was not the first who formulate the basics of a modern research university; his ideas still describe the essences of research universities (see more at Backhous, 2015, pp. 5–11; Menand, Reitter, & Wellmon, 2017, pp. 3–4): the aim of a university is to teach students to think through the integration of research and teaching; the university exists for the sake of knowledge; research plays the main role in increasing quality of teaching content as well as the academic freedom of the university must be safeguarded from external influences. Von Humboldt viewed science "not as something already found but as knowledge that will never be fully discovered and, yet, needs to be searched for unceasingly transformed the idea of the university from a training institution primarily devoted to the transmission of established knowledge to the revolutionary concept of an institution primarily committed to the search for, the pursuit of, the discovery of, knowledge" (Muller, 1985). Von Humboldt has seen the universities as "space where logical and critical thinking could be complemented with the tasks of transmitting knowledge" (Montesinos et al., 2008). Finally, "the difference between a school teacher and a university professor are the teacher shall instruct pupils and the professor shall cooperate with students to serve aspiration for science and research" (Bachhaus, 2015, p. 58).

Two interdependent concepts—the Third Mission and the Triple Helix underline roles of universities in relations to society, industry and governance (Zawdie, 2010). Even those, understanding of the Third Mission concept varies in different contexts (see Jaeger & Kopper, 2014), according to Pausits (2015), the third mission bundles all activities outside the academic environment and promotes interaction with stakeholders typically in lifelong education and technology transfer and innovation. Montesinos et al. (2008) elaborated diverse dimensions in third mission as:

- social third mission (e.g. services to society with no economic benefit).
- enterprising third mission (e.g. service related to technology transfer and patents, consulting for industry and governments, providing advisory services, networking with entrepreneurs).
- innovative third mission (e.g. business networking for patent exploitation, joint ventures with industrial sectors, research for innovation in specific industries or companies).

The Triple Helix concept underline role of universities in knowledge society trough their interaction industry, and government. The Triple Helix concept, underline that (Etzkowitz, 2003): networks among university, industry, and governments lead to innovation rather than work of anyone alone; the role of universities in developing that kind of networks are crucial and the role of universities in developing countries can be important in helping the local economy to grow through technology "catch-up".

2.2 Internationalization

Focus of many HEIs is on internationalization. Rationales for increasing focus of HEIs on internationalization can be based on building reputation of HEIs in order to be more successful than others in attracting international students and in market share of educational and research services (Knight, 2004). According to Knight (2004) internationalization is "the process of integrating an international, intercultural or global dimension into the purpose, functions or delivery of post-secondary education." Forms of internationalization can be different and classified as *at home* and *abroad. Internationalization abroad* is based on "all forms of education across borders of HEIs" (Knight, 2004), e.g. students mobility as well as "cross-border delivery of education, internationalization of the curriculum, internationalization of teaching and learning, and internationalization of learning outcomes" (Wulz & Rainer, 2015). *Internationalization at home* is based on educational activities in HEIs like use of international literature and teaching case studies based on international cases, lectures provided by visiting professors and industry representatives (Beelen & Jones, 2015).

Internationalization of curriculum is "the purposeful integration of international and intercultural dimensions into the formal and informal curriculum for all students within domestic learning environments" (Beelen & Jones, 2015). In those definition term of formal curricula is related to syllabus related to planed activities for achieving course aims and learning outcomes, and informal curricula is related to extracurricular activities which are not objects of assessments but contribute to achieving course aims.

2.3 Nontraditional Providers of Higher Education

Capacities of many universities are not sufficient to respond to intensive diversification and internationalization of labor market as well as increased needs for a specialization of potential employees. The labor market is moving faster than higher education can respond (Goldman Sachs Global Investment Research, 2015). "Given the increase in demand for higher education, there are new providers, delivery methods, and types of programs. As a result, there are new types of higher education providers active in the delivery of education programs both domestically and internationally" (Knight, 2004). According to Goldman Sachs Global Investment Research (2015, p. 10), most obvious threat to traditional higher education is based on potential employers' change of a attitude toward nontraditional providers of higher education. According to the same source, 30% of undergraduates already take some classes online (mostly MOOCs) and offers of non-degree programs where curricula are designed in accordance to the best global companies are booming. Furthermore, increasing numbers of companies are using new systems for talent identification (e.g. GitHub) as a better indicator than student's academic resume.

2.4 Post Millenials and Generation Z

At colleges and universities, all around the globe, we have a new generation—Generation Z, with unique values, priorities, and intention as compared to Millennials (Fatemi, 2018). New generations of students require significant changes in higher education due their specific mindset. It is quite difficult to say clearly their age on the global level, in North America, Western Europe or Asian developed countries it can be said that Generation Z are cohort born roughly after 1995. Studies based on cases from developing countries describe the Generation Z as global-generation that we can meet now globally (see more in Dabija, 2018).

While Millenials were called digital natives, Generation Z represents the first generation to have lived entirely in a digital era and have had access to more information than any other generation (Marron, 2015). Their digital abilities and need to be connected online constantly created generations for whom "online word-of-mouth matters greatly" (Goldman Sachs, 2015, pp. 11–12). Generation Z accepts social media much earlier in their life comparing with Millenials. Socializations of Generation Z is based on social media and "this phenomenon may cause problems with social interactions and conflict resolution at college, work … Mentoring may be increasingly important for Gen Z to overcome social problems" (Marron, 2015). Marron (2015) suggests that reverse mentoring (e.g. experienced and older executives are mentored by younger employees on specific topics) can be valuable for socializing Generation Z.

Generation Z values diversity and they want to be unique, to establish their specific identities, to express their different opinions, to be engaged in collaborative problems solving, to design "third spaces" that allow interactions outside departmental boundaries or to promote dialogue, create identify and build community (Goldman Sachs, 2015, pp. 11–12). Generation Zs are financially conservative and entrepreneurial.

Based on studies that include more than 150,000 students, Seemiller and Grace (2017) noticed that learning for Generation Z students is different from learning of Millenials and highlight some features of new Generation Z students such as:

- they prefer to be engaged in learning in which they can immediately apply what they learn to real life;
- they prefer to process information first trough observing—they like to watch others complete tasks before applying the learning themselves (seeking information through video);
- they want to know that learning content they are learning have broader applicability to more than just a practice example;
- they are accustomed to learning independently;
- they view peers and instructors as valuable resources and like to have the option to work with others on their own terms;
- they believe that practical experiences like internships are essential in a college education, gain connections, experience, and skills that they can leverage in any future occupation.

Seemiller and Grace (2017) suggested that in order to effectively engage Generation Z students, the essential is to: utilize video-based learning and voice-over instructional videos; incorporate learning independently into class and group work; offer adequate engagement opportunities for students trough case study competitions and hack-athons and provide options for internship opportunities. Regarding to study of Jermyn (2018) the concentration or attention issues of generation Z, due overwhelming with information, can be solved by "microlearning"—learning in little bits. Students from Generation Z, want to know how they can use or benefit from the knowledge or skills they were learned.

3 How to Teach About Standardization to Enhance Learning to Generation Z

3.1 Teaching Is About Interactions

It should be clear that education about standardization is not only teaching about standardization. There is extensive literature which seeks to capture the complexity of education, teaching, and learning in modern higher education. In the context of this study, "teaching refers to all forms of interaction with students that relate to their academic experience in order to enhance student learning" (Fry, Ketteridge, & Marshall, 2015, p. 8). A teaching is only one of many different ways to support learning (Ackoff & Greenberg, 2008, p. 4).

Learning theories stress the importance of interactivity in learning and call for creative engagement of students. In the context of education, interactivity means students' active participation in the learning process, instead of their passive reproduction of the content provided by teachers (O'Connell, 2007). Interactivity is related to creative engagement, where engagement is defined as "student-faculty interaction, peer-to-peer collaboration, and active learning …" (Chen, Gonyea, & Kuh, 2008). Moore (1989), Anderson and Garrison (1998) and Pask (1976) define three types of interaction:

- student–content interaction,
- student–teacher interaction, and
- student–student interaction.

According to Johnson and Johnson (1989), learning tends to be the most effective when students work together towards a group solution to the given problem, work collaboratively, express their thoughts, discuss and challenge the ideas of others. Interactions among students facilitate the development of critical thinking skills, skills of self-reflection and co-construction of knowledge and meaning (Brindley, Walti, & Blaschke, 2009). Siemens (2002) defined four levels of student-student interaction: communication (talking, discussing); collaboration (sharing ideas and working together in a loose environment); cooperation (doing things together, but each one with his/her own purpose) and community (striving for a common purpose).

Generation Z students needs at least some sort of digital interaction in their university education (Morris, 2014). According to CONE Generation Z CSR Study (2017, p. 17): "To reach Generation Z through online content, the key is to make it as visual and engaging as possible"—this generation is far more likely to want to see videos, pictures or stories. In the context of online interactive support materials for standardization course, and because of insufficient capacities of HEIs, cooperation among universities and nontraditional providers of higher education with support of interested stakeholders is necessary.

Contents related to standardization and standards can be more interesting for Generation Z students, than it was for Millenials, because they are entrepreneurial and willing to be involved in solving "big problems". The understanding standardization is important in modern economy and business. "Standards control access to virtually every market in global commerce and directly affect more than eighty percent of world trade ... Standards are being used to support sustainability purposes and their use may further increase" (Purcell & Kushnier, 2016). Global and knowledge-based economy influences establishments of new mechanisms of standards development (e.g. cooperation and/or consortia based standardization). Development of standards and standardization in many different areas, subjects and levels and substantive influence of standards development organizations—make standardization more important and interesting to learn, teach and research about it. Generation Z students are more likely to want to see written articles or graphical presentation, than ex-cathedra lecture, furthermore, long articles and data-heavy content increases in effectiveness as Generation Z students mature (CONE Generation Z CSR Study, 2017, p. 17). Taking into account unique features of Generation Z students, content related to standardization and standards should be:

- Engaging as possible. Instead of examples and success stories related to standards and standardization, Generation Z students need to be questioned and provoked to express their different opinions. Due their developed ability to learn independently (Seemiller & Grace, 2017), reading assignment should be followed by class discussion.
- Based on real-world. Generation Z prefers to be involved in meaningful activities and real-world problem-solving. Generation Z students need to be faced with real-world problems in a way to facilitate their involvement. More real-worlds problems and practice related to standardization and standards are needed.
- Truly global. Generation Z is interested to face with social and global problems more than Millenials. In that context, more independent research and adequate case studies related to effectiveness of standards and standardization globally are needed.

In that context, role of SDOs and other organizations that develop or support developing teaching material related to standardization is important. However, first it is needed to understand that the aim of a university is to teach students to think, teachers role is to enhance students learning and critical thinking, but overall educational experience that incorporate practical learning opportunities in standardization depends much on fruitful cooperation among universities, SDOs, industry and others actors. This process of cooperation can be more connected with activities of universities related internationalization *at home* and *abroad*.

3.2 Teaching Methods Are Based on Wanted Learning Outcomes

In the context of education, it is important to understand first what learning outcomes are wanted. Among some standardization professionals there is a belief that, once standardization is recognized as a regular subject in higher professional and university curricula, general awareness and appreciation of standards' benefits will automatically result (Simons, 1999). Furthermore, Simons (1999) noted that—"masters of standardization are not needed" it is better to enlarge the group of people that are aware of the usefulness of standardization and acquaint possible future decision-makers with the subject. However, learning outcomes of education about standardization in higher education can be significantly higher than only raising awareness.

According to Bloom's Taxonomy (Bloom, 1994) skills in the cognitive domain revolve around knowledge, comprehension and critical thinking of a particular topic. According to Bloom's Taxonomy (1994), in general, three different types of learning outcomes can be identified, i.e. acquisition of factual knowledge, application of the acquired knowledge, and analysis, synthesis and evaluation of knowledge (Fig. 1).

Factual knowledge has the shortest span of usability as it is based on memorizing or remembering. In the digital era, with high dynamics of data and information, focusing on remembering as learning domain, is highly unreliable. Ex cathedra lecturing (in a classroom, in a blended learning environment (integration online digital media with traditional classroom teaching) or online) can provide only short span learning outcomes. If the acquisition of factual knowledge is desired, then interaction with learning content is the most influential (Mijatovic, Jovanovic, & Jednak, 2012). According to Seemiller and Grace (2017) study, Generation Z students are accustomed to individual learning—interaction with learning content by themselves (e.g. reading assignments) is important for Generation Z to prepare them for application of knowledge.

If higher levels of learning outcomes are to be achieved, more sophisticated teaching and learning techniques have to be used. Higher levels of learning outcomes need more communication and two-way interaction. Application of knowledge is

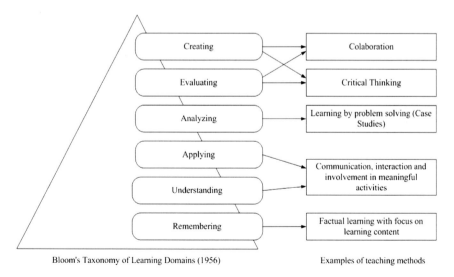

Fig. 1 Methods for teaching are based on what kind of learning outcomes are wanted

based on the understanding of teaching content. Application of knowledge can be achieved through learning by doing, practical assignment, adequate internships, and field trips, connections with professionals (e.g. mentoring by professionals when working on assignments). Furthermore, involvement in more challenging activities (e.g. problem-solving, case studies), in order to reach more demanding learning objectives (like analysis, synthesis, and evaluation) increase students' need for student-teacher and student-student interaction.

Methods for teaching about standardization are based on what kinds of learning outcomes are wanted. Based on Bloom's Taxonomy (1956) and possible standardization learning outcomes (based on the study of Blind & Dreshler, 2017, p. 76), teaching methods that can be used to enhance students learning about standardization at four levels were showed in Fig. 2. If wanted learning outcomes are:

- informing students about standardization and standards,
- familiarizing students with standardization and standards or
- obtaining awareness of students about impacts of standardization or standards implementation,

in educational theory, that teaching outcomes might be identified as factual knowledge. In this case, lack of extensive theory background might reject university teachers into include standardization into their curricula. Nevertheless, factual knowledge is important and serves as a base for higher levels of knowledge, factual knowledge

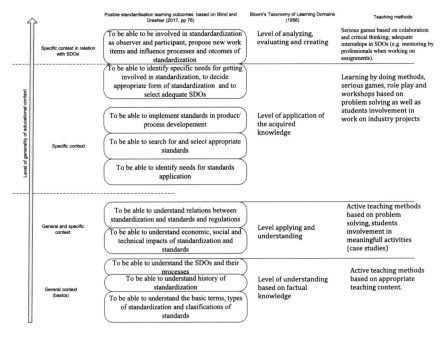

Fig. 2 Teaching methods based on possible standardization learning outcomes

in standardization, is not enough to have prepared Generation Z students to use their knowledge in their future carriers.

However, Generation Z students are more likely to choose courses which provide practical and applied knowledge. In standardization, as in many others disciplines, application of knowledge is related to specific context (e.g. it is more difficult to address general effectiveness of standards than effectiveness of specific standards or group of standards). The teaching methods depend on levels of generality of educational content in the meaning that teaching general (e.g. related to all disciplines) and specific aspects of standardization needs different teaching methods (Fig. 2).

3.3 First Step: Group Discussions

Teaching standardization based on group discussions in class, blended learning environment or online should be carefully designed to provoke interaction with learning content, among students and with teachers. Study of Rezaei (2017) found that six major factors play significant roles in the effectiveness of group discussion: mode of instruction (face-to-face vs. online), type of task (convergent vs divergent), the anonymity of participants in on/line environment, homogeneity in skill levels of students, peer assessment, and group size.

According to Vass and Littleton (2010), there are different kinds of group discussions (disputional, cumulative and exploratory). Disputional discussion is characterized by disagreement, lack of cooperation in decision making and the lack of clear resolution. Cumulative group discussion is characterized by some extent of positive but uncritical exchange among group members. Exploratory group discussion is based "on critical and reflective negotiation of different views among group member, constructive criticism, identification reason for assertion as well as counter reasons and alternatives" (Vass and Littleton, 2010). Exploratory discussions have the highest educational value (Krčadinac, Jovanović, & Devedžić, 2012). In the context of education about standardization, special focus should be on exploratory class discussion as a tool for development of students' collaboration skills needed for standardization practice (e.g. negotiation, consensus building, decision-making and dealing with conflict).

3.4 Bringing Practice into a Classroom

Bringing practice into a classroom can be done in many ways. The study of Katusic et al. (2017) acknowledges the need that students work with standards before they graduate, and reports positive experiences with student involvement in combination with student projects, research activities and industrial projects with a focus on application of new standardized technologies, as the case of enabling students to get their own experience in applying telecommunication standards.

Topics related to practical details of standard development (e.g. meaning of consensus in voluntary standardization; development of standards in consortia or standards battles) can be explained by teaching methods based on management simulation games and role playing in classes. In a blended and online environment, serious games (games designed for learning purposes rather than entertainment) can be used. In principle, there are many companies able to develop games needed for university education about standardization, for the right price. Such educational games must be designed in accordance to adequately defined learning outcomes. The experience gained with Setting Standards, a one-day simulation exercise on strategy and cooperation in standardization developed by the Delft University of Technology and United Knowledge, sets an example (http://setting-standards.com).

One of the key take away from NIST's Industry-Academic Teaching Support Workshop 2015 (NIST, 2015, p. 9) was that case studies are needed to teach about standardization and standards. The case study method was developed by Harvard University, more than a hundred years ago, with the aim to expose business students to 'real' corporate problems (Credle, Beale, & Maheshwari, 2009). The teaching case study is a written description of a real situation or event, in which a person or people in an organization are faced with decisions, challenges, opportunities, problems or attitudes (Erskine, Leenders, & Mauffette-Leenders, 1998). A teaching case is not a success story, example or task with only one correct solution. To crack (deeply understand) and solve a case, as well as to present their solution and recommendation, students need to search for data and information, work in teams (usually of four students), and apply their knowledge in a specific context. A teaching case study can be short (2–3 pages) for working on class (90 min), medium (up to 10 pages) for use as a course assignment (e.g. team of four have one or two weeks to crack, solve and prepare their presentation or report) or long (more than 20 pages) for a case study competition. In the context of education about standardization, a significant part of the teaching material should be based on teaching case studies in order to confront students with real problems, challenges and opportunities in standardization.

In higher education case study method can be used in teaching (as a part of formal curricula) and for case study competitions (as a part of informal curricula). International business case study competitions are the great place for signaling and identification of talents. There are many case studies in international students' competition which address more or less problems related to standardization. The world's largest and most diverse collection of resources for business education worldwide (http://www.thecasecentre.org)—has more than 100 case studies with keyword standardization. Based on the study of Damnjanovic, Proud and Ruangwanit (2016), Damnjanović and Mijatović (2017) student benefits from being engaged in international case study competition are numerous: unique learning experience, increased self-confidence, better creative thinking, higher understanding of a multidisciplinary approach, familiarity with multicultural environment, development of business logic, team building skills, improvement of managerial skills (presentation skills, teamwork skills, work under pressure), analytical thinking, stepping out of factual learning and understanding the broader context, and networking. Moreover,

Teaching Standardization to Generation Z-Learning ...

international case study competitions can be seen as an area of exchange of experience in teaching and learning practice.

Many university teachers might find it quite cumbersome to teach Generation Z using only traditional methods. For many university teachers, particularly in developing countries, teaching methods incorporating students' active participation (e.g., working on case studies, class discussions, role play, collaborative problem solving, case study competitions, etc.) are new and innovative ways of teaching (Almarghani & Mijatovic, 2017; Bollag's, 1996). Internationalization *abroad* and *at home* can be valuable to improve skills of teachers to adequately teach to Generation Z students. Teachers' mobility, visiting professorships, joint research projects and academic programs can be tools for dissemination of education about standardization and internationalization of curriculum related to standardization.

Standardization needs researchers; more precisely standardization needs academic researchers too. In order to have standards based on relevant and up-to-date scientific and technological knowledge, the involvement of researchers in the standardization process is necessary (Blind & Gauch, 2009). Results of the study of Mijatovic, Krstic, & Komazec (2013), showed that only 36% of academics involved in work of national technical committees in Serbia transfer information about standards and standardization to their students. Having at least some practical experience in standardization is not sufficient motivator for academics to teach about standardization. Higher education has it owns rules, academic walls and specific issues and support to university teachers and researchers to research and publish about standardization can help them to establish deeper involvement in the education about standardization (Mijatovic, Horvat, & Krsmanovic, 2014). More adequate and available teaching content in form of high quality case studies, serious games, videos, texts and stories customized to satisfied learning preferences of Generation Z students are needed.

4 Conclusion

According to estimation of Fatemi (2018), Generation Z will represent 75% of the workforce by 2030, at least half of them will work on positions where they will need certain understanding and knowledge of standards and standardization in the specific context. Where and how will they acquire needed knowledge and skills will shape at least part of our future. To reach Generation Z, in a classroom, a blended learning environment or online—adequate teaching based on interactivity student-content; student-teacher and student-student is of highest importance. Even Web-based teaching materials and online courses without possibilities for certain quality of interaction with students will be insufficiently attractive to generation Z.

Many calls, initiatives, and activities in support of higher education about standardization aim to enlarge the group of people that are aware of the usefulness of standardization and acquaint potential future decision-makers with standardization. However, the learning outcome of higher education about standardization should be significantly more ambitious than 'raising awareness'. Moreover, if the goal is merely to promote standardization, there are plenty of places other than universities or higher education to do so. Based on Bloom's Taxonomy (1956), teaching methods that can be used to enhance Generation Z students learning about standardization can be systematized at four levels of acquired knowledge:

- understanding is based on factual knowledge which can be achieved in the general educational context (e.g. all disciplines). To enhance these learning outcomes, active teaching methods based on appropriate teaching content followed by exploratory class discussion, can be an adequate choice.
- applying and understanding which needs some level of specialization (e.g. understanding of economic and social impacts of standards should be based on specific (types, groups of) standards). To enhance these learning outcomes active teaching methods based on problem-solving (case studies, students assignments based on real problem solving).
- practical application of acquired knowledge needs a specific educational context. To enhance these learning outcomes, learning by doing methods based on serious games, role play, as well as student involvement in industry/research projects would be right choice.
- analyzing, evaluating and creating which needs specific educational as well as standardization context. To enhance this learning outcome, vise choice can be serious games based on collaboration and critical thinking; adequate internships in SDOs (e.g. mentoring by professionals when working on assignments).

Many steps forward have been made in education about standardization recently, but it seems that they are isolated from one another and have not taken into account the previous work of others. In Europe, fruitful cooperation among governments, industry, academia, and SDOs is needed in order to well-establish and nurture education about standardization. This cooperation needs to be based on understanding key roles universities can and should play. Academic freedom needs to be adequately understood and guarded in calls and initiatives for education about standardization. Cooperation among universities, nontraditional providers of higher education, governments, industry and SDOs related education about standardization can be more closed to increasing efforts of many universities related to internationalization and third mission development.

References

Ackoff, R. L., & Greenberg, D. (2008). *Turning learning right side up: Putting education back on track*. Pearson Education Inc.

Almarghani, E. M., & Mijatovic, I. (2017). Factors affecting student engagement in HEIs—It is all about good teaching. *Teaching in Higher Education, 22*(8), 940–956. https://doi.org/10.1080/13562517.2017.1319808.

Anderson, T. & Garrison, D. R. (1998). Learning in a networked world: New roles and responsibilities. In C. Gibson (Ed.), *Distance Learners in Higher Education*. Atwood Publishing.

Backhaus, J. G. (2015). *The university according to Humbolt—History, policy, and future possibilities. Springer briefs in economics.* https://doi.org/10.1007/978-3-319-13856/5.

Beelen, J., & Jones, E. (2015). Redefining internationalization at home. In A. Curaj, L. Matei, R. Pricopie, J. Salmi, & P. Scott (Eds.), *The European higher education area between critical reflections and future policies* (pp. 59–72). Heidelberg: Springer International Publishing. https://doi.org/10.1007/978-3-319-20877-0_5.

Blind, K., & Gauch, S. (2009). Research and standardisation in nanotechnology: Evidence from Germany. *J Technol Transfer, 34,* 320–342. https://doi.org/10.1007/s10961-008-9089-8.

Blind, K. & Dreshler, S. (2017). European market needs for education in standardisation/standardisation-related competence. *Directorate-General for Internal Market, Industry, Entrepreneurship and SMEs (European Commission).* https://doi.org/10.2873/540387.

Bloom, B. S. (1956). Taxonomy of educational objectives. *The Cognitive Domain.* New York: David McKay.

Bloom, B. S. (1994). Reflections on the development and use of the taxonomy. In L. W. Anderson, & L. A. Sosniak (Eds.), *Bloom's taxonomy: A forty-year retrospective.* Chicago National Society for the Study of Education.

Bollag, B. (1996). Reform efforts appear stalled at colleges in Eastern Europe. *The Chronicle of Higher Education, 43*(8), A-59.

Brindley, J. E., Walti, C., & Blaschke, L. M. (2009). Creating effective collaborative learning groups in an online environment. *The International Review of Research in Open and Distance Learning, 10*(3), Retrieved from: http://www.irrodl.org/index.php/irrodl/article/view/675/1271.

Chen, P., Gonyea, R., & Kuh, G. (2008). Learning at a distance: Engaged or not? *Innovate, 4*(3). Retrieved from: http://innovateonline.info/pdf/vol4_issue3/Learning_at_a_Distance-__Engaged_or_Not_.pdf.

Choi, D. G., & De Vries, H. J. (2011). Standardization as emerging content in technology education at all levels of education. *International Journal of Technology and Design Education, 21*(1), 111–135. https://doi.org/10.1007/s10798-009-9110-z.

Choi, D., & de Vries H. J. (2013). Integrating standardization into engineering education: The case of forerunner Korea. *International Journal of Technology and Design Education, 23*:1111–1126. https://doi.org/10.1007/s10798-012-9231-7.

Choung, J. Y., Ji, I., & Tahir, H., (2011) International standardization strategies of latecomers: The cases of Korean Tpeg, T-Dmb, and binary CDMA. *World Development, 39*(5), 824–838.

CONE GEN Y CSR STUDY: How to speak with Z. (2017). Retrieved 21.05.2018 from http://www.conecomm.com/2017-cone-gen-z-csr-study-pdf/.

Credle, S. H., Beale, P. L., & Maheshwari, S. (2009). The use of case analysis training and competitions to assure learning and school-wide quality. *Business Education & Accreditation, 1*(1), 29–44. https://doi.org/10.12691/jbe-5-1-2.

Dabija, D. C. (2018). Enhancing green loyalty towards apparel retail stores: A cross-generational analysis on an emerging market. *Journal of Open Innovation: Technology, Market, and Complexity, 4,* 8. https://doi.org/10.1186/s40852-018-0090-7.

Damnjanović, V., & Mijatović, I. (2017). Student perception of benefits from being engaged in international case study competitions management. *Journal of Sustainable Business and Management Solutions in Emerging Economies, 22*(2). https://doi.org/10.7595/management.fon.2017.0017.

Damnjanovic, V., Proud, B., & Ruangwanit, N. (2016).Perceived benefits and issues of student learning in business case competition—Comparison study of Serbia, Australia and Thailand. *Athens Journal of Education,* forthcoming issue, Retrieved 23 May 2017 from https://www.athensjournals.gr/education/2017-1-X-YDamnjanovic.pdf, https://doi.org/10.1080/03797720802254072.

de Vries, H. J. (2015). How to implement standardization education in a country. In K. Jakobs (Ed.), *Modern trends surrounding information technology standards and standardization within organizations* (pp. 262–275). Hershey, PA: IGI Global. https://doi.org/10.4018/978-1-4666-6332-9.ch015.

Erskine, J. A., Leenders, M. R., & Mauffette-Leenders, L. A. (1998). *Teaching with cases*. London: Ivey Publishing.

Etzkowitz, H. (2003). Innovation in innovation: The triple helix of university-industry-government relations. *Social Science Information, 42*(3), 293–337, SAGE Publications.

European Commission. (2011). Communication from the Commission to the European Parliament, The Council, The European Economic and Social Committee and The Committee of the Regions. *Supporting Growth and Jobs—An Agenda for the Modernisation of Europe's Higher Education Systems—COM/2011/0567 final*. Retrieved 12.3.2018 from https://eur-lex.europa.eu/legal-content/EN/TXT/HTML/?uri=CELEX:52011DC0567&from=EN.

Fatemi, F. (2018). *What's your strategy for attracting generation Z?*. Entrepreneurs #LikeABoss, March 31, 2018. Retrieved 21.05.2018 from https://www.forbes.com/sites/falonfatemi/2018/03/31/whats-your-strategy-for-attracting-generation-z/#7cdd0a56cad4.

Forbes Coaches Council. (2019). *Here comes Gen Z: How to attract and retain the workforce's newest generation*. https://www.forbes.com/sites/forbescoachescouncil/2018/02/27/here-comes-gen-z-how-to-attract-and-retain-the-workforces-newest-generation/2/#52fd064f1849.

Fry, H., Ketteridge, S., Marshall, S. (2015). *A handbook for teaching and learning in higher education: Enhancing academic practice*. Routledge.

Goldman Sachs Global Investment Research. (2015). *Emerging Theme Radar: What if I told you ... themes, dreams and flying machines*. Retrieved 12.05.2018 from http://www.goldmansachs.com/our-thinking/pages/macroeconomic-insights-folder/what-if-i-told-you/report.pdf.

Greisler, P. (2013). Welcome addresses. In B. Henningsen, J. Schlaeger, & H. E. Tenorth (Eds.). *Humboldt's Model: The Future of the Universities in the World of Research: Conference Report*. BWV Verlag.

Jaeger, A., Kopper, J. (2014). Third mission potential in higher education: Measuring the regional focus of different types of HEIs. [Jahrbuch für Regionalwissenschaft], *Review of Regional Research, 34*(2), 95–118. Retrieved 1.4.2018 from https://EconPapers.repec.org/RePEc:spr:jahrfr:v:34:y:2014:i:2:p:95–118.

Jermyn, D. (2018). How colleges are adapting for the new Gen Z. *Special to the Globe and Mail*, 1.6.2018. https://www.theglobeandmail.com/business/careers/business-education/article-how-colleges-are-adapting-for-the-new-gen-z/.

Johnson, D. W., & Johnson, R. T. (1989). *Cooperation and competition: Theory and research*. Edina, MN: Interaction Book Company.

Katusic, D., Skocir, P., Kusek, M., Jezic, G., Ratti, C., & Bojic, I. (2017). Hands-on education about standardization: Is that what industry expects? *IEEE Communications Magazine, 55*(5), 138–144. https://doi.org/10.1109/MCOM.2017.1601134.

Knight, J. (2004). Internationalization remodeled: Definition, approaches, and rationales. *Journal of Studies in International Education, 8*(1), 5–31. https://doi.org/10.1177/1028315303260832.

Krčadinac, U., Jovanović, J., & Devedžić, V. (2012) Visualizing the affective structure of students interaction. In S. K. S. Cheung, J. Fong, L. F. Kwok., K. Li, R. Kwan (Eds.), *Hybrid learning. ICHL 2012. Lecture notes in computer science* (Vol. 7411). Berlin, Heidelberg: Springer.

Light, G., Cox, R., Calkins, S. (2009). *Learning and teaching in higher education: The reflective professional* (pp. 38–43). SAGE Publication.

Marron, M. B. (2015). New generations require changes beyond the digital. *Journalism & Mass Communication Educator, 70*(2), 123–124. https://doi.org/10.1177/1077695815588912.

Menand, L., Reitter, P., & Wellmon, C. (Eds.). (2017). *The rise of the research university a sourcebook*. The University of Chicago Press.

Mijatovic, I., Horvat, A., Krsmanovic, M., (2014). Academics' and researchers' participation in the National Technical Committees in Serbia. *EURAS Proceedings 2014, Cooperation Between Standardization Organizations and the Scientific and Academic Community* (pp. 135–149). Germany: Wissenschaftsverlag Mainz GmBH Aachen. ISBN: 978-3-86073-305-2.

Mijatovic, I., Jovanovic, J., & Jednak, S. (2012). Students online interaction in a blended learning environment—A case study of the first experience in using an LMS. In *Proceedings of the*

4th International Conference on Computer Supported Education (ESEeL-2012) (pp. 445–454). https://doi.org/10.5220/0003963804450454, ISBN: 978-989-8565-07-5.

Mijatovic, I., Krstic, I., & Komazec, S. (2013). Experiences with participation of Serbian academics in national technical committees. In *Proceedings of the 18th EURAS Annual Standardisation Conference "Standards: Boosting European Competitiveness", Brussels, Belgium*, June 24–26, 2013 (pp. 279–295). ISBN: 978-3-86130-655-9.

Montesinos, P., Carot, J. M., Martinez, J. M., & Mora, F. (2008). Third mission ranking for world class universities: Beyond teaching and research. *Higher Education in Europe, 33*(2/3). ISSN 0379-7724 print/ISSN 1469-8358 online/08/02/30259-13 # 2008 UNESCO.

Moore, M. G. (1989). Editorial: Three types of interaction. *American Journal of Distance Education, 3*(2), 1–7.

Morris, L. V. (2014). Editor's page: Who's listening? *Innovative Higher Education, 39*:1–2. https://doi.org/10.1007/s10755-014-9283-6.

Muller, S. (1985). Wilhelm Von Humboldt and the University in the United States. *Johns Hopkins APL Technical Digest, 6*(3), 253–256. Retrieved 12.3.2018 from http://www.jhuapl.edu/techdigest/views/pdfs/V06_N3_1985/V6_N3_1985_Muller.pdf.

NIST National Institute of Standards and Technology. (2015). Industry-Academic teaching support. *Workshop*–Summary Report—July 2015.

O'Connell, J. (2007). *Creative Web 2.0 learning. A talk given at Christian Teacher Librarians Association conference, Sydney*, May 23, 2007. Retrieved from: http://www.slideshare.net/heyjudeonline/creative-web-20-learning.

Pask, G. (1976). *Conversation theory: Applications in education and epistemology*. Amsterdam: Elsevier.

Pausits, A. (2015). The Knowledge Society and Diversification of Higher Education: From the Social Contract to the Mission of Universities. In A. Curaj, L. Matei, R. Pricopie, J. Salmi, & P. Scott (Eds.), *The European Higher Education Area: Between Critical Reflections and Future Policies* (pp. 267–284). https://doi.org/10.1007/978-3-319-20877-0_18.

Purcell, D., & Kushnier, G. (2016, Mart/April). Globalization and standardization. *The Journal of SES—The Society for Standards Professionals*. Retrieved from http://www.standardsuniversity.org/e-magazine/august-2016-volume-6/globalization-and-standardization/.

Rezaei, A. (2017). Features of successful group work in online and physical courses. *The Journal of Effective Teaching, 17*(3), 5–22, Retrieved 12.05.2018. from https://www.uncw.edu/jet/articles/Vol17_3/Rezaei.pdf.

Seemiller, C., Grace, M. (2017). Generation Z: Educating and engaging the next generation of students. *About Campus*. https://doi.org/10.1002/abc.21293.

Siemens, G. (2002). Interaction. *E-learning course*, October 8, 2002. Retrieved May 19, 2008, from http://www.elearnspace.org/Articles/Interaction.htm.

Simons, C. A. J. (1999). Education in standardization—Getting structured common sense into our society—The personal opinion of standards educator. *ISO Bulletin*.

Unterhalter, E., & Carpentier, V. (2010). *Global inequalities and higher education: Whose interests are you serving?* Palgrave Macmillan.

Vass, K., Littleton, K. (2010). Peer collaboration and learning in the classroom. In K. Littleton, C. Wood, J. K. Staarman (Eds.), *International handbook of educational psychology* (pp. 112, 113). Emerald Group Publishing Limited.

Verman, L. C. (1973). *Standardization*. Archon Books.

Wulz, J., & Rainer, F. (2015). Challenges of student mobility in a cosmopolitan Europe. In A. Curaj, L. Matei, R. Pricopie, J. Salmi, & P. Scott (Eds.), *The European higher education area between critical reflections and future policies* (pp. 43–58). Heidelberg: Springer International Publishing.

Zawdie, G. (2010). Knowledge exchange and the third mission of universities : Introduction: The triple helix and the third mission—Schumpeter revisited. *Industry and Higher Education, 24*(3), 151–155. https://doi.org/10.5367/000000010791657437.

Ivana Mijatović is associate professor at Faculty of Organizational Sciences, University of Belgrade. She is passionate teacher; on bachelor studies she teaches Standardization 1, Quality Engineering and Quality Planning, on master and Ph.D. studies she teaches Standardization 2 and ICT Standardization. In 2018/2019, she is the Chair of the Board of the International Cooperation for Education about Standardization ICES (http://www.standards-education.org/). Since 2017, she has been serving as a member of working group related to EU Joint Initiative on Standardization (JIS Action 3). She serves as a vice-president on the board of the European Academy for Standardization (EURAS, http://www.euras.org) and she was a member of the Balkan Coordination Committee for Standardization, Protypes and Quality (BCC). She is a member of national technical committee KS I1/07 - Software engineering, IT for Education and Internet at Institute for Standardization of Serbia (national mirror committee in relation with ISO/IEC JTC 1/SC 7; ISO/IEC JTC 1/SC 36 and ISO/IEC JTC 1/SC 40 IT Service Management and IT Governance). She developed course Standardization 1 and wrote the text book with 10 case studies Standardization 1 (2015). Her current academic work addresses the question of standardization, education about standardization, teaching quality and quality aspects of technology enhanced learning.

Summary

Sustainable Development: Knowledge and Education About Standardisation—Discussion

Henk J. de Vries

This book aims to bridge the worlds of sustainability, standardisation and education. Unfortunately, the majority of contributions we received, does not address sustainability but the few that do this show convincingly that the three belong together. Starting from the sustainability side—in the transition towards a more sustainable future, relying on governments is not sufficient anymore. Industry will not take the lead either, and NGOs may want to but lack power. The three sides have to join forces and setting common agreements is an important part of this: standards. Standards are needed to specify what needs to be done or to provide performance criteria, and to provide methods to test if requirements are being met. From a government perspective, these standards are an alternative to legislation, or can be used in combination with regulation. The voluntary character may be seen as a weakness but are a strength as well if stakeholder involvement leads to real commitment. Starting from the business side: the attention for sustainability is shifting from mitigating negative externalities towards creating new business opportunities by providing products and services that help solve societal problems. For NGOs the challenge is not just to create alternatives for the small percentage of consumers really interested in sustainability but to team up with industry and governments to make changes that have much more societal impact.

Many stakeholders know little about standards and standardisation. This applies to governments and societal stakeholders, but also industry experts lack proper knowledge—both at strategic and at tactical and operational levels. In this book, this market need for education about standardisation is addressed by Blind and Drechsler (2018). They investigated the manifest need at the business side. Apparently the companies in their dataset are not yet aware of the bridge between sustainability and standardisation. More in general, the latent need for education is not being addressed.

H. J. de Vries (✉)
Rotterdam School of Management, Erasmus University, Rotterdam, The Netherlands
e-mail: hvries@rsm.nl

Faculty of Technology, Policy and Management, Delft University of Technology, Delft, The Netherlands

© Springer Nature Switzerland AG 2020
S. O. Idowu et al. (eds.), *Sustainable Development*, CSR, Sustainability,
Ethics & Governance, https://doi.org/10.1007/978-3-030-28715-3_13

Kanevskaia adds this dimension. She talked with standardisation experts and discovered that their expertise was one-sided—they could not put their standards-related work into perspective. She concludes that a multidisciplinary approach to standardisation education is needed.

In the study by Blind and Drechsler (2018), more than one third of the respondents reports that standardisation belongs to their job responsibilities with the first year of employment and for almost another one third within the second to fifth year. This underpins the need to include standardisation in formal education, more specific knowledge can be included in vocational training. This book focuses on formal education at university level.

The first part provides several examples of teaching about standardisation. The differences between these chapters are striking. If curricula were designed starting from the market need perspective, one might expect more coherence, even independent from the discipline—as argued by Kanevskaia: a multidisciplinary approach is needed anyhow. UN-ECE did an attempt to develop such a common curriculum, see the chapter by Jachia et al. The Korean initiative to develop an ISO Workshop Agreement on Education about Standardisation even goes one step further—getting standards for professionals in the field of standardisation. As De Vries et al. argue in their chapter, building on Kuhn (1962): this is a normal step in the development of scientific fields, where the development of research, teaching and the establishment of scientific communities go hand in hand.

Back to these differences between the teaching stories in Part 1: it seems that the individual knowledge and experiences are decisive in the design of a course. In particular Fomin expresses this in a very personal way. All researchers who share their experiences in these chapters make use of their own research. Van de Kaa shows that this also may go the other way round: students contributing to research. Not only research experience is being used, some scholars extend this with other forms of standardisation experience. Van den Bossche, for instance, makes use of his involvement in European standardisation activities and makes students experience such work via the assignments he gives them.

The Bulgarian case (Vasoleva) shows that also national policies and even involvement of governments and/or standards bodies may impact education about standardisation. The new Romanian standardisation policy includes education about standardisation, the answers Romanian respondents gave to Puiu suggest a positive attitude towards inclusion of standardisation in education, despite the biases that may be inherent to the research method he used. Bulgaria and Romania are neighbouring countries and share a history of a state economy under communist leadership, a revolution in 1989 and European Union membership since 1987. Nevertheless the two papers show major differences and this suggests that addition of other country papers would have added even much more diversity. The cases of van den Bossche (Belgium) and de Vries (The Netherlands) show that also national standards bodies may stimulate education about standardisation, in different ways. De Vries, Manders, and Veurink (2014) show more ways to do this.

The dependencies on experiences of individual teachers are also a cause of the fragmentation in education about standardisation described by Jachia, and the lack of coherence in curricula. Apparently the feasibility of standardised curricula depends not only on the demand side but also on the supply side. In most countries, university professors are free in what they teach.

An unexpected element in even the majority of chapters is the effort teachers spend in making their teaching attractive to students. All cases described in the first part of the book show examples and some of them are very creative. So the diversity applies not only to contents but also to teaching methods. The cases have in common that students get actively involved and Mijatovic argues in her chapter that this is needed for the current generation of students, Generation Z. The importance of attractiveness in teaching about standardisation was noticed before (De Vries & Egyedi, 2007) but then more emphasis was on teaching materials. Our cases suggest that methods are even more important than materials but of course both are needed and these are strongly related. Mijatovic emphasizes that students want to be challenged and engaged—challenging them to provide solutions to sustainability problems is probably appealing to them. She provides guidance for teaching methods related to possible standardisation learning outcomes defined by Blind and Dreschler but unfortunately those outcomes do not include sustainability—they relate too much to 'old', technical standardisation.

The paper by De Vries shows that a systematic design of a multidisciplinary standardisation course automatically leads to the inclusion of sustainability aspects, in his case even to addressing almost all United Nations Sustainable Development Goals. However, this does not imply that sustainability as such is taught in a systematic way. Wright et al. start at the sustainability side and include standardisation. As far as we know, the Master in "Standardization, Social Regulation and Sustainable Development" at the University of Geneva is the only one in the world that combines the two in a balanced way. For this programme, the university cooperates with the International Organization for Standardization (ISO), teaching materials are available to other universities via ISO member bodies: national standards institutes.

To conclude, standardisation is essential to make sustainability happen and for both concepts, better education is needed. This may start at the standardisation side or at the sustainability side, or combine both from scratch. To make this happen, a combination of top-down and bottom-up approaches is needed (CEN, CENELEC, & ETSI, 2012; De Vries, 2012). Our chapters focus on bottom-up, the Bulgarian case shows a combination. This book puts emphasis on the importance of attractiveness of teaching to make it appealing to generation Z. This relates to the art of teaching by individual teachers but may be supported at national or international level by developing approaches and related materials such as serious games, see Mijatovic' chapter.

References

Blind, K., & Drechsler, S. (2018). *European market needs for education in standardization/standardization-related competence*. Luxembourg: Publications Office of the European Union.

CEN, CENELEC, & ETSI. (2012). *Masterplan on education about standardization*. Brussels: CEN CENELEC Management Centre. https://www.cencenelec.eu/standards/Education/JointWorkingGroup/Documents/Masterplan%20on%20Education%20about%20Standardization.pdf.

De Vries, H. J. (2012). *Implementing standardization education at the national level*. In K. Jakobs (Ed.), *Innovations in organizational IT specification and standards development* (pp. 116–128). Hershey, PA: IGI Global. https://dx.doi.org/10.4018/978-1-4666-2160-2.ch006.

De Vries, H. J., & Egyedi, T. M. (2007). Education about standardization—Recent findings. *International Journal of IT Standards and Standardization Research, 5*(2), 1–16. https://doi.org/10.4018/jitsr.2007070101.

De Vries, H. J., Manders, B., & Veurink, J. (2014). *Cooperation between national standards bodies and universities*. Rotterdam: Rotterdam School of Management, Erasmus University.

Kuhn, T. S. (1962). *The structure of scientific revolutions*. Chicago: University of Chicago Press.

Henk J. de Vries (1957) is Professor of Standardisation Management at the Rotterdam School of Management, Erasmus University in Rotterdam, The Netherlands, Department of Technology and Operations Management, Section Innovation Management, and Visiting Professor at Delft University of Technology, Faculty of Technology, Policy and Management, Department of Values, Technology and Innovation, Section Economics of Technology and Innovation. His education and research focus on standardisation from a business point of view. From 1994 until 2003, Henk worked with NEN, Netherlands Standardization Institute, in several jobs, being responsible for R&D during the last period. Since 1994, he has an appointment at the Erasmus University's School of Management and since 2004, he has been working full-time at this university. Henk is (co-)author of more than 380 publications on standardisation, including several books. See www.rsm.nl/people/henk-de-vries. In 2009, the International Organization for Standardization ISO awarded his education about standardisation as best in the world. Henk is President of the European Academy for Standardisation EURAS.

Index

A

Academic curriculum, 101, 103–105, 107, 162, 165
Academic programme about standardization, 90, 91
Acceptance (of standard), 83
Active learning, 5, 6, 197
Active teaching methods, 204
Ageing population, 67
American Society for Quality (ASQ), 187
Applications developers, 37
Asia Pacific Economic Cooperation (APEC), 18, 71, 146, 153, 185, 187
ASME B16.4, 161
ASRO, 102, 103, 106, 107
Awareness, 6, 34, 95, 96, 98–102, 107, 108, 113, 114, 141, 144, 152–155, 162, 164, 165, 169–171, 174, 185, 192, 199, 200, 204
Awareness-raising on standards education, 100, 144, 152, 199, 204

B

Balkan Committee for Standardization, Protyping and Quality, 180, 182
Batteries, 22, 23, 25, 26, 37
Belt and Road University Alliance of Standardization (B&RUAS), 180, 182, 192
Benefits of using standards, 9, 26, 45, 87, 95, 102, 114, 119, 152, 153
British Standards Institution, 8, 99
British Standards Online, 100
Bruface, 17, 18
Bulgaria, xii, 79–87, 89, 92, 146, 180, 212

Bulgarian Institute for Standardization (BIS), 82, 83, 85
Bulgarian State Standards (BDS), 81–84
Bulgaria's accession to the European Union, 82
Business, 4, 5, 9, 12, 13, 18, 39, 40, 47, 57–61, 64, 67, 68, 70, 71, 86–88, 95, 99, 100, 102, 113–115, 119, 125, 127, 128, 134, 135, 140–143, 145, 146, 148, 150, 151, 154, 155, 162, 163, 170, 172, 173, 181–183, 185, 194, 198, 202, 211
Business administration, 127
Business knowledge and skills, 125
Business school, 58, 59, 71, 88
Business sector, 61, 62

C

Capacity building in teaching standards, 152
Carbon footprint, xii, 8, 140
Case study, 7–12, 23, 51, 52, 85, 196, 202, 203 management, 9, 85, 91, 202
CEN, 64, 81, 82, 95, 98, 99, 102, 103, 149, 150, 173, 184, 213
CENELEC, 18, 20, 22, 64, 82, 95, 98, 99, 102, 103, 149, 150, 173, 184
Central and Eastern European countries, 81, 82
Certification, 67, 69, 81, 84, 86–88, 90, 91, 117, 119, 128, 147, 148, 170, 186, 187
Charging, 19, 21–23, 25, 26
Charging accessories, 25
Charging station, 22
Chronology of education about standardization in Bulgaria, 80
Circular economy, 64
Cities, 68, 71
Classroom, 5, 6, 9, 11, 91, 199, 201, 203

© Springer Nature Switzerland AG 2020
S. O. Idowu et al. (eds.), *Sustainable Development*, CSR, Sustainability, Ethics & Governance, https://doi.org/10.1007/978-3-030-28715-3

215

216   Index

Climate, 3, 8, 12, 57, 66, 68, 69
Communication, 7, 19, 22, 35, 49, 63, 91, 102,
    113, 114, 126, 133, 164, 168, 197, 199
Community of practice, 180, 181
Companies perspective, 121
Competence demands, 129
Competences in standards, 114, 117, 119, 120,
    125, 130, 131, 133, 134
Competences of employees, 116, 128
Competences skills, 26, 87, 113–117, 119, 120,
    125, 135
Competency model, 129, 133
Consensus Bluetooth, 167, 168
Consensus process, 169
Consortia, 33, 34, 115, 123, 128, 130, 161,
    198, 202
Consumer, xi, 4, 25, 36, 49, 61, 62, 67, 69, 80,
    82, 84, 85, 92, 140, 164, 184, 211
Consumer choices, 38
Consumption, 68, 79, 140
Cooperation in teaching standards, 202
Copolco (Committee on Consumer Policy),
    184
Cradle to Grave, 9
Cross-sectoral standardization, 162, 174
Curriculum, 5, 7, 17, 18, 21, 57, 59, 61–65, 71,
    83, 85, 86, 89, 132, 146, 149, 162, 165,
    173, 187, 194, 195, 203, 212
Curriculum design, 57, 59, 62, 71
Curriculum design education, 57, 59, 62, 71
Cyber-human systems, 41

D
Danish Standards Foundation, 101
De facto rules, 167
De facto standards, 33, 34, 37, 40
De jure standards, 33, 34
Dilemmas (of management), 63, 65
Dimensions (of standarisation), 61
DIN EN 2501-1, 161
Discipline, 34, 35, 59, 60, 63, 81, 86, 89, 96,
    129, 135, 153, 162–165, 167, 169,
    171–175, 179, 180, 182, 185, 191, 201,
    204, 212
Dominant designs, 48
Dooyeweerd, 59, 63
Drinking water, 68

E
Economic growth, 66, 68, 191
Economic policy, 162
Ecosystem, 69
Education, xi, xii, 3, 5–7, 10, 11, 13, 19,
    45–47, 51–53, 57–62, 66, 67, 70, 79,
    81, 83, 85–91, 95–101, 105, 107, 108,
    114, 115, 118, 120, 128, 130–132, 134,
    135, 142, 144–155, 165, 171–174,
    179–183, 185–187, 191, 192, 194–198,
    202, 211–213
Education about standardization, 35, 40–42,
    79–86, 89, 92, 95–99, 102, 103, 149,
    150, 162, 165, 173, 174, 180, 191–193,
    197, 199, 201–204
Education about standardization in Europe, 98,
    100, 150, 182
Educational and training programmes on
    standardization, 150
Educational program, 34, 35, 101, 102, 162,
    165, 170, 172, 173, 175
Education in Romania, 102
Education on metrology, 128, 145, 147, 148,
    151, 155
Education on standardization, 46, 47, 52, 145,
    150, 151, 154, 171
Education on Standards and Standards-Related
    issues ("StaRT-ED") Group, 150
E-learning, 100, 102, 147
Electric vehicle, xii, 17, 18, 21–23
Employability, 5, 13
Employees, 39, 67, 95, 99, 101, 103, 105, 113,
    115, 116, 118–120, 124, 126–129, 133,
    143, 153, 154, 163, 173, 186,
    195, 196
Employees working in standardization, 126,
    129
Employment, 5, 51, 68, 82, 129, 171, 212
Energy, 8, 9, 18, 21, 23, 49, 62, 68, 70, 140,
    145
Engineer, xi, 17, 71, 115, 117, 119, 125, 143,
    144, 153, 162, 166, 171, 172
Engineering, 17–21, 61, 97, 102, 103,
    127–129, 134, 145, 147–149, 153–155,
    162, 163, 174, 175, 179, 182
Environmental, 3–5, 7, 9, 10, 12, 18, 66, 69,
    70, 84, 86, 87, 141, 149, 163
Ergonomics, 67
Ethical norms, 41
European Academy for Standardisation
    (EURAS), 31, 99, 150, 180, 182
European Commission, 19, 22, 41, 95, 114,
    120, 161, 167, 168, 182, 191, 192
European market needs for education in
    standardisation, 137
European Standardization Organization, 95, 98,
    99, 108, 149, 150
European standardization policy, 79
European standards, 19, 80,
    82, 83, 92, 168

Index

European Union (EU), 20, 22, 67, 79, 82, 84, 87, 98, 100, 149, 165, 167, 168, 172, 185, 187, 191, 212

F

Factors, 8, 10, 17, 48–53, 83, 101, 153, 168, 181, 183, 201
Fairness, 169
Fair, Reasonable and Non-Discriminatory (FRAND), 166, 167
Financial reporting, 163, 167
4th industrial revolution, The, 40
Forest, 69
Formal standardisation, 64, 115, 184
Formats, 7, 11, 33, 37, 60, 173

G

Gender equality, 67
General competence model for standardization, 114
Generation Z, xii, 193, 195–199, 201, 203, 204, 213
Germany, 83, 99, 100, 126, 143, 144, 146, 182, 185
Globalization, 92, 97, 170, 174, 183, 184
Good practices, 95, 103–105, 107, 192
Governance, 64, 79, 80, 146, 161, 164, 170, 172, 194
Government, xi, 3, 19, 33, 34, 62, 66, 68, 69, 85, 96, 101, 102, 107, 150, 154, 155, 174, 185, 187, 191, 194, 204, 211, 212
Graduate skills, 4, 5
Greece, 180
Guided inquiry, 53

H

Higher education, 5, 81–83, 85–87, 89, 91, 96, 99–105, 107, 133, 134, 145, 149, 165, 171, 191–193, 195, 197, 199, 202–204
Higher education system, 98
High level structure (of management system standards), 69
History of standards, 102
Human interface, 60, 65, 67

I

Impact, 9, 10, 12, 18, 20, 23, 26, 62, 64, 65, 67–69, 79, 84, 86, 99, 130–132, 140, 141, 156, 161, 167, 169, 170, 184, 187, 200, 204, 211, 212
Implementation (of standard), 19, 117, 119, 144, 165, 166, 173, 184, 200
Importance of teaching standards, 144
Inclusiveness, 67–69, 162, 169

Incorporating standards-related issues into educational curricula, 150, 153
Indonesia, 47, 68, 145, 146, 184, 192
Industrialisation, 32, 68
Industry 4.0, 40
Inequality, 3, 57, 66, 68
Informatics, xii, 31, 34–37, 40
Information and communication technologies, 33, 161
Information systems, xii, 31, 148
Infrastructure, 18, 21, 22, 37, 68, 69, 82, 131, 132, 146, 155, 164, 165
International Federation of Standards users IFAN, The, xii, 119, 142, 143, 183–185, 187
Innovation, 8, 20, 34, 57–60, 62–65, 67–71, 113, 124, 141, 162, 164, 185, 187, 194
Innovation management, xii, 58–60, 64, 70, 71, 117
Inquiry based teaching, 45–47, 49, 53 factors, 49, 53
Institutions, 19, 47, 52, 69, 81–83, 85–87, 91, 92, 96, 99, 101, 107, 128, 130, 131, 144–150, 152–154, 162, 170, 172, 174, 191–193
Interaction, 19, 20, 22, 124, 131, 132, 161, 181, 194, 196, 197, 199–201, 203
Inter-disciplinarity, 42, 164, 165, 171
Interface, 36–38, 40, 58–60, 65, 67, 68, 70, 71, 164, 185
International Cooperation for Education about Standardization (ICES), 95, 99, 120, 150, 180, 187, 192
International Electrotechnical Commission (IEC), 18–20, 22–24, 26, 31, 39, 64, 103, 115, 120, 140, 149, 169, 173, 184
International Federation of Standards users (IFAN), xii, 142, 143, 183–185, 187
Internationalization, 193–195, 198, 203, 204
International Labor Organization (ILO), 163
Internet of Things, 167
Internet technologies, 36–38
Interoperability, 4, 11, 36–38, 40, 59, 141, 163, 169, 170
Intriguing strategy, 42
ISO, xi, 4, 8, 9, 12, 19, 20, 22, 23, 26, 33, 39, 64, 81, 85–90, 96, 98, 101, 103, 108, 113, 115, 126, 140, 149, 153, 169, 173, 184, 186, 212, 213
ISO 14001, 69, 86–88, 163
ISO 9001, 67, 87, 88, 163, 187
ISO Award "Higher Educationin Standardization", 85, 89
IT- knowledge and skills, 125

J

Japan, 67, 96, 145, 146
Job descriptions in relation to standardization, 116
Joint Initiative on Standardization (JIS), 114, 182
Joint Initiative on Standardization - Action 3, 191
Justice, 69, 168
JWG-EaS, 98, 99, 101

K

Knowledge, xi, xii, 4–8, 10, 12, 17, 31, 32, 34, 36–40, 42, 47, 50–52, 59, 60, 64, 65, 84, 85, 89–92, 96–102, 105–107, 113–121, 125, 127–135, 142–144, 147, 151–153, 165, 171–174, 179–183, 185, 191, 193, 194, 196–204, 211, 212
Knowledge about standardization, 97
Korea (Republic of), 185
Kuhn, 180, 212

L

Learning, 5, 6, 9–13, 36, 37, 42, 46, 50, 52, 60, 67, 89, 90, 97, 102–104, 151, 152, 165, 171, 174, 191–204
Learning outcomes, 26, 32, 38, 193–195, 198–200, 202, 204, 213
Lectures, 5, 11, 18, 19, 21, 45, 46, 60, 65, 91, 146, 172–174, 195
Legal pluralism, 163
Level of education, 105, 106, 126
Life cycle analysis, 69
Life Cycle Assessment (LCA), 9, 10, 12, 87, 88
Life-long learning, 99

M

Maintaining the standard, 36
Malaysia, 185
Management, xi, xii, 4, 9, 19, 31, 39, 40, 45, 47, 57–60, 63–71, 80, 86–89, 91, 92, 102, 113, 117, 118, 123, 125–129, 134, 143, 144, 146, 149, 155, 162, 164–167, 170, 172, 175, 179, 183, 186, 202
Management of technology, 47
Management system standards, 39, 156
Mandatory standards, 33, 80, 84, 92
Market access, 164, 167
Market demand for knowledge on standards, 114, 142
Masterplan for education, 98
Master's course, 19
Matching problem, 58, 59, 63

Measurement standard, 59, 62
Moot court, 173
Multi-disciplinarity, 144, 162, 164, 165, 170, 172–174
Multi-disciplinary, 171, 172
Multidisciplinary character of standards, 154
Multi-disciplinary education, 162, 165, 174, 175
Multimedia, 36–38

N

National committee, 25
National, regional and international standards, 11
National standards body, 7, 19, 35, 60, 96–99, 101, 103, 106, 145, 155, 212
National technical committees, 203
Need for standard, 9, 11, 18, 130, 185
Netherlands, 47, 58, 69, 99, 100, 147, 212
Network, 22, 23, 33, 38, 48, 49, 68, 86–88, 100–102, 107, 128, 164, 167, 173, 180, 181, 194
Network effect, 38, 48, 49, 68, 164
Network externalities, 38, 167
Network of Practice, 23
Non-conformance, 39, 40
Non-Governmental Organisation (NGO), xi, 4, 68, 103, 107, 185, 211
Nontraditional providers of higher education, 193, 195, 197, 204

O

Ocean, 69
Openness, 166, 168, 169
Open source, 167, 169
Organizational management, 34, 39, 40
Organizational norm, 39, 40
Organizational studies, 39
OSI interoperability model, 33

P

Participation in standardization processes, 113, 117
Path dependency, 39, 49
Peace, 69
Pedagogies, 5, 6, 12, 150, 192
Plastic soup, 69
Platforms, 64, 65, 68, 98, 100, 103, 125, 150, 168, 174, 184, 186
Platform wars, 55
Post-communist transformation, 82
Poverty, 57, 66, 67
Practical exercises, 18–21, 26
Practical relevance, 45–48, 52

Index 219

Practice, 4, 5, 10, 12, 19, 20, 23, 32, 33, 39, 40,
47, 57, 59–61, 64, 71, 81, 86, 87, 89, 92,
98, 100, 131, 144, 151, 155, 172, 173,
180, 183, 184, 196, 198, 201, 203
Practitioner, xii, 5, 47, 48, 50, 66, 165, 173,
175, 180, 183
Private voluntary standards, 167, 170
Production, 10, 68, 79, 80, 84, 119, 127, 140,
141, 143, 152, 183
Profession, 17, 153, 154, 171, 180, 183,
185–187
Professional, 5, 26, 61, 87, 96–98, 107, 117,
118, 120, 126–129, 134, 143, 144, 154,
162, 165, 170, 171, 173–175, 179–185,
187, 191, 198, 200, 204, 212
Professional development, 5
Public good, 162

Q

Quality management, 39, 61, 69–71, 81,
85–87, 117, 129, 134, 147, 148, 152,
163, 183, 187

R

R&D, 117, 124, 128, 129, 167–169
Real-world, 6–8, 10, 11, 171, 198
Real-world learning, 60
Reflection, 6, 10, 197
Regulation, 3, 4, 18, 20, 26, 57, 79, 81, 117,
143, 146, 163, 164, 168, 172, 211, 213
Religion, 70
Research, 3, 7, 17, 18, 21, 34, 35, 42, 45–53,
59–64, 67, 68, 70, 71, 87, 91, 97–103,
106, 113, 116, 119–121, 128, 132, 134,
135, 143, 144, 150, 161, 162, 164–167,
170, 171, 174, 180, 182, 183, 187,
193–195, 198, 201, 203, 204, 212
Responsible standardization, 41
Role of standards, 8, 11, 12, 31, 36, 67–69, 85,
96, 97, 101–103, 108, 123
Role-playing, 6
Routine, 39, 40, 140
Royalty rate, 166
Royalty-free, 166

S

Schumpeter, 59
Scope of standardisation, xi
SDOs, 119
Sea, 69
Serbia, 147, 191, 203
Simulation, 5–7, 10–12, 100, 101, 147, 173,
202

Skills, 4–7, 12, 17, 26, 45, 47, 85, 87, 89–91,
103, 113–121, 125–127, 129–131,
133–135, 142–144, 150–153, 171, 174,
181, 183, 191, 192, 196, 197, 199,
201–203
Socialist era, 79, 80, 92
Society for Standards and Standardization
(SSS), 185
Stakeholder, xi, 4, 33, 49, 57, 60–62, 65, 67,
70, 71, 82, 84, 85, 92, 96, 101, 102, 107,
149, 150, 154, 161, 162, 164, 166–170,
182, 185, 192, 194, 197, 211
Standard, xi, xii, 4–12, 18–20, 22, 23, 25, 26,
31–42, 48–53, 57–59, 61–65, 67–70,
81, 84, 95, 115, 117–119, 121, 122,
132–135, 143, 162, 165, 166, 169, 170,
173, 179, 184, 187
Standard approval, 84
Standard development, 49, 115, 128, 202
Standard generalist, 118
Standard implementation, 115, 119
Standard operating procedure, 40
Standard specialist, 118
Standard-Developer, 116
Standardization, xi, xii, 4, 7–9, 11, 57–65,
67–71, 89–91, 121, 123, 147, 179–187,
211–213
Standardization task, 186, 187
Standardization, 113–120, 125–135
Standardization and companies, 156
Standardization and Regulatory Techniques
("STaRT") Group, 151
Standardization bodies and organizations, 19
Standardization during centrally planned
economy, 83
Standardization education, 18, 45, 83, 89, 95,
100, 101, 165, 170, 171, 192
Standardization in Europe, 98, 100, 114, 150
Standardization learning outcomes, 200
Standardization-related competencies, 91
Standards and innovation, 101, 141
Standards and international trade, 34, 64, 65,
140, 156
Standards and sustainable development, xi
Standards as barrier to trade
Standards as regulatory tools, 153
Standards battles, 45, 46, 48–50, 53, 202
Standards competition, 36
Standards course, 96–98, 100–107
Standards Development Organizations (SDOs),
115, 117, 148, 191, 192, 198, 204
Standards drafting process., 18
Standards-Engineer, 116

Standards Engineering Society (SES), 180
Standards-Manager, 116
Standards in Romania, 97, 98, 102, 103, 105–107, 212
Standards of Council for Mutual Economic Assistance (CMEA standards), 81
Standards professional, 144, 174, 179, 183, 184, 186
Standards-related courses, 87, 88, 147
  education on metrology, 147
Standards users organisation, 180, 182–184, 186
Strategy on standardization, 107
Structured inquiry, 45, 46, 50, 52, 53
Structured-inquiry-based teaching, 53
Student, 5–12, 17–19, 23, 25, 26, 31, 32, 35–38, 40–42, 45–53, 58–60, 62, 64, 65, 67–71, 85, 86, 89, 91, 95, 97, 99–101, 103–107, 114, 132, 144–147, 151, 152, 154, 162, 165, 170–173, 193–198, 200–204, 212, 213
Sustainability, xi, xii, 3–5, 7, 10, 11, 57, 58, 66–68, 70, 71, 87–89, 155, 166, 191, 198, 211, 213
Sustainable development, 3, 4, 6–8, 10–12, 57, 66, 69, 79, 86, 89, 92, 139, 143, 144, 146, 153, 155, 213
Sustainable Development Goals (SDG), 3, 7, 57, 66, 67, 69–71, 155

T
Teaching, xii, 6, 31, 32, 35–42, 45–49, 51–53, 58, 59, 70, 83, 86, 87, 89, 91, 95–97, 101, 107, 151–154, 165, 172, 192–195, 197–204, 212, 213
  education, 6, 35, 40–42, 45, 47, 51, 53, 70, 83, 86, 89, 91, 95–97, 101, 107, 144, 146, 147, 151, 152, 154, 165, 192–194, 197–199, 201–204, 212, 213
Teaching about standardization, 40, 193, 197, 200
Teaching about standards and standardization, xii, 41
Teaching case study, 202
Teaching method, 32, 49, 198
Teaching standardization, 89, 144, 146, 147, 150, 151, 153, 154, 201
Technical barriers to trade, 141, 144, 155, 167
Technical committee, 18, 22, 103, 203
Technical knowledge and skills, 125
Technical regulations, 144, 155, 167
Technical specifications, 35, 161, 163, 166

Technical standards, 32, 39, 163, 165, 166, 170
Technological innovation, 97, 167
Technology, xi, 7, 11, 17–22, 26, 34, 36–38, 40, 41, 45, 47, 48, 50, 51, 53, 58, 60, 62, 81, 83, 85, 88, 100, 102, 117, 134, 135, 141, 144–149, 161, 164, 166, 170, 172, 191, 192, 194, 201, 202
Technology management, 172
Telecommunications, 32, 49, 95, 102, 127–129, 134, 140, 149, 166, 167, 170, 201
Third Mission, The, 194
Trade association, 61, 184, 185
Training materials on standardization, 154
Transatlantic Trade and Investment Protocol (TTIP), 67, 168
Triple Helix, The, 194

U
Understanding of voluntary standards, 141, 170
UNECE model educational programme on stanardization, 151
UNECE Recommendation "I", 150
UNECE Working Party on Regulatory Cooperation and Stanardization Policies (Working Party 6), 144, 150–152, 155
Uniform quality standards (CMEA standards), 81
United Nations, xi, xii, 3, 66, 139, 155, 213
United Nations Economic Commission for Europe (UNECE), xii, 18, 20, 139, 140, 142, 144, 146, 150–156, 191
United Nations Sustainable Development Goals, xi, 57, 67, 71, 213
University, 5, 7, 8, 17, 19, 31, 34, 41, 45–47, 53, 57, 58, 60, 68, 71, 79, 83, 85–90, 92, 95–102, 107, 114, 116, 126, 129, 134, 143–155, 171, 172, 174, 180, 185, 191–195, 197–200, 202–204, 212, 213

V
Van de Lagemaat, 59, 60, 63
Verman, 61, 96, 179, 191
Von Humboldt, 193
Vrije Universiteit Brussel, 17–19, 21

W
Water, 68, 69
Web Hypertext Application Technology Working Group, 169
Web protocols, 164

Index 221

Wi-Fi, 163, 166
Winner takes all, 48
Wireless Local Area Network, 163

Wood, 69
Working group, 98, 99, 102, 149, 150
World Wide Web Consortium, 168

Printed in the United States
By Bookmasters